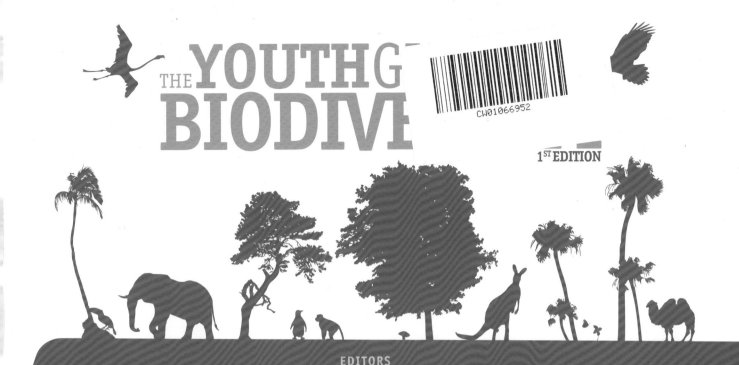

THE YOUTH GUIDE TO BIODIVERSITY

1ST EDITION

EDITORS
Christine Gibb :: Neil Pratt :: Reuben Sessa

AUTHORS
:: David Ainsworth :: Nadine Azzu :: Dominique Bikaba :: Daniel J. Bisaccio :: Kate Buchanan Zeynep :: Bilgi Bulus ::
:: David Coates :: Jennifer Corriero :: Carlos L. de la Rosa :: Amanda Dobson :: Maria Vinje Dodson :: Cary Fowler ::
:: Christine Gibb :: Jacquie Grekin :: Caroline Hattam :: Terence Hay-Edie :: Saadia Iqbal :: Leslie Ann Jose-Castillo ::
:: Marie Aminata Khan :: Conor Kretsch :: Ping-Ya Lee :: Charlotte Lusty :: Michael Leveille :: Claudia Lewis ::
:: Ulrika Nilsson :: Kieran Noonan-Mooney :: Kathryn Pintus :: Neil Pratt :: Ruth Raymond :: John Scott :: Reuben Sessa ::
:: Junko Shimura :: Ariela Summit :: Giulia Tiddens :: Tamara Van 't Wout :: Jaime Webbe ::

THE YOUTH GUIDE TO BIODIVERSITY

The designations employed and the presentation of material in this information product do not imply the expression of any opinion whatsoever on the part of the Food and Agriculture Organization of the United Nations (FAO) concerning the legal or development status of any country, territory, city or area or of its authorities, or concerning the delimitation of its frontiers or boundaries. The mention of specific companies or products of manufacturers, whether or not these have been patented, does not imply that these have been endorsed or recommended by FAO in preference to others of a similar nature that are not mentioned.

The views expressed in this information product are those of the author(s) and do not necessarily reflect the views of FAO.

ISBN 978-92-5-107445-9

All rights reserved. FAO encourages reproduction and dissemination of material in this information product. Non-commercial uses will be authorized free of charge, upon request. Reproduction for resale or other commercial purposes, including educational purposes, may incur fees. Applications for permission to reproduce or disseminate FAO copyright materials, and all queries concerning rights and licences, should be addressed by e-mail to copyright@fao.org or to the Chief, Publishing Policy and Support Branch, Office of Knowledge Exchange, Research and Extension, FAO, Viale delle Terme di Caracalla, 00153 Rome, Italy.

© FAO 2013

United Nations Decade on Biodiversity

A United Nations Decade on Biodiversity product.

This document has been financed by the Swedish International Development Cooperation Agency, Sida. Sida does not necessarily share the views expressed in this material. Responsibility of its contents rests entirely with the authors.

FRONT COVER PHOTO:
© Diego Adrados (age 16)

BACK COVER PHOTOS:
© Reuben Sessa :: © FAO/Alessia Pierdomenico :: © Mila Zinkkova/Wikimedia Commons :: © Danil Nenashev/World Bank

YOUTH AND UNITED NATIONS GLOBAL ALLIANCE

Acknowledgements

The Youth and United Nations Global Alliance (YUNGA) would like to thank all the authors, contributors, graphics designers and other individuals and institutions who have supported the development of this guide. All have found the extra time in their busy schedules to write, edit, prepare or review the materials, and many others have kindly allowed the use of their photos or other materials. Special thanks go to the staff at the Food and Agriculture Organization of the United Nations (FAO) and the Secretariat of the Convention of Biological Diversity (CBD) for the time and effort they dedicated in the preparation of this guide, in particular to Diana Remarche Cerda for drafting the original outline and Alashiya Gordes and Sarah Mc Lusky for reviewing the text. Deep gratitude also extends to Studio Bartoleschi for the endless patience in revising and updating the layout and graphics. All the contributors have a great passion for biodiversity and hope that the guide will inspire young people to learn, better understand the importance of biodiversity and take action in conservation initiatives. In addition, gratitude goes to the YUNGA and CBD Ambassadors for their passion and energy in promoting this guide.

THE YOUTH GUIDE TO BIODIVERSITY

TABLE OF CONTENTS

ACKNOWLEDGEMENTS
page iii

PREFACE
page vi

1 WHAT IS BIODIVERSITY
page 1

2 HOW ARE PEOPLE AFFECTING BIODIVERSITY?
page 13

3 GENES AND GENETIC DIVERSITY
page 23

4 SPECIES: THE CORNERSTONE OF BIODIVESITY
page 39

5 ECOSYSTEMS AND ECOSYSTEM SERVICES
page 57

6 TERRESTRIAL BIODIVERSITY - LAND AHOY!
page 71

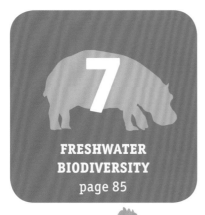
7 FRESHWATER BIODIVERSITY
page 85

8 THE RICHES OF THE SEAS
page 103

9 IN FARMERS' FIELDS: BIODIVERSITY AND AGRICULTURE
page 117

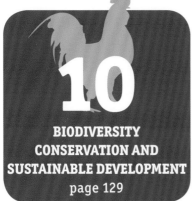
10 BIODIVERSITY CONSERVATION AND SUSTAINABLE DEVELOPMENT
page 129

11 BIODIVERSITY AND PEOPLE
page 145

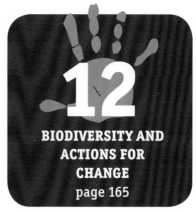
12 BIODIVERSITY AND ACTIONS FOR CHANGE
page 165

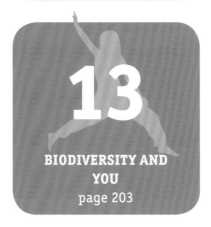
13 BIODIVERSITY AND YOU
page 203

ANNEXES
A
Contributors & organisations
page 225
B
List of species
page 235

GLOSSARY
page 239

THE YOUTH GUIDE TO BIODIVERSITY

PREFACE

A CAMOUFLAGED CREATURE THAT CAN LOOK IN TWO DIFFERENT DIRECTIONS AT THE SAME TIME...

A CREEPY CRAWLY WITH NO EYELIDS...

A HUGE, TRUMPETING MONSTER THAT CAN SMELL WATER FROM A DISTANCE OF THREE MILES (4.8KM)...

NO, THIS IS NOT A DESCRIPTION OF A TOTALLY WILD SCIENCE FICTION MOVIE!

"Welcome to planet Earth, whose inhabitants include chameleons, who can see in two different directions at the same time, insects without EYELIDS and elephants with their great sense of smell.

These are just a few examples. The variety of animals and plants on Earth is truly wondrous. Moreover, its diverse ecosystems, such as deserts, oceans, rivers, mountains, marshlands, forests, and grassy plains are specifically suited to the creatures and plants that live there. But changes to an ecosystem's environment can spell doom for its native plants and animals, and unfortunately, this is happening all too fast today. Many species are at risk of disappearing entirely. While extinction has always happened as a natural part of a gradual evolutionary process, the current rate of extinction of animals and plants is thought to be hundreds, perhaps even thousands, of times faster than that brought about by natural evolutionary processes.

Biodiversity experts say that nowadays most extinctions are caused by human activity, such as deforestation, mining, conversion of land, building dams, roads and cities, overfishing, and other activities that lead to habitat destruction, climate change, and pollution. So much so that the International Union for Conservation of Nature (IUCN) has 5 689 entries on its endangered species list (**www.iucnredlist.org**), many of whom might be familiar to you like species of gorillas, orangutans, turtles, eagles, whales, cranes,

seals, foxes, bears, and tigers but also many species of plants, birds, insects, reptiles, amphibians and fish.

Most of us believe that all life has the right to exist, and many of us also feel a personal loss when wildlife is damaged or destroyed. However, loss of Earth's biodiversity affects us in material ways too. In fact, biodiversity is the foundation on which human life depends. Plants and animals provide food and medicine, rivers provide precious drinking water, and trees absorb greenhouse gases and protect land from erosion. Damaging natural ecosystems may also affect natural processes, such as flood control and crop pollination, among others.

We invite you to dive into this comprehensive youth guide for in-depth insights into biodiversity, the benefits it provides to us, the threats it faces, and what actions we can take to protect it. The guide is richly illustrated, including award-winning photos taken by youth from around the world as part of the 'See the Bigger Picture' contest that supported *The Green Wave*, a global campaign promoting biodiversity. At the end of the guide there is a useful tool for setting up an action plan and undertaking your own biodiversity project, with Six Simple Steps towards Change. Take inspiration from far-reaching projects of other young global leaders and their innovative projects. At the end of each chapter and in the annexes, you will find additional resources, assignments for further learning about your surroundings, and other useful information.

> **IT MIGHT BE TOO LATE TO SAVE SOME SPECIES FROM EXTINCTION, BUT IT'S NOT TOO LATE TO TAKE ACTION TO SAVE OTHERS. PEOPLE LIKE YOU CAN MAKE THE BIGGEST DIFFERENCE, AND GETTING INFORMED AND MOTIVATED IS A GREAT WAY TO START.**

Anggun, Jean Lemire, Carl Lewis, Fanny Lu, Debi Nova, Lea Salonga & Valentina Vezzali

THE YOUTH GUIDE TO BIODIVERSITY

CBD, FAO & YUNGA AMBASSADORS

ANGGUN
YUNGA AND FAO GOODWILL AMBASSADOR

"Humanity must learn to share the planet with other species and as individuals we must change our daily habits to help preserve our biodiversity."

JEAN LEMIRE
CBD AMBASSADOR

"The health of our planet relies on an exquisitely delicate balance and extraordinary diversity of life. The more we learn about biodiversity and discover its beauty, the more we care for it."

DEBI NOVA
YUNGA AMBASSADOR

"May this guide inspires you to experience and explore the wonders of nature, preserve it and motivate your family, friends, classmates, and community to save our planet's biodiversity."

YOUTH AND UNITED NATIONS GLOBAL ALLIANCE

FANNY LU
YUNGA AND FAO GOODWILL AMBASSADOR

"I hope this guide opens your eyes to the incredible biodiversity around us and motivate you to take action."

CARL LEWIS
YUNGA AND FAO GOODWILL AMBASSADOR

"We must win the race against time to preserve what we have left of our biodiversity, every action that you and I take is important."

LEA SALONGA
YUNGA AND FAO GOODWILL AMBASSADOR

"Our world is truly wonderful; let us learn to live in harmony with it and preserve it for future generations to also enjoy."

VALENTINA VEZZALI
YUNGA AMBASSADOR

"We're surrounded by awe-inspiring and life-enriching plants and animals. Can you imagine living in a world without them? I can't – so let's take a stand for biodiversity!"

Download this guide and other interesting resources from:
www.yunga.org

WHAT IS BIODIVERSITY?

DEFINING BIODIVERSITY AND ITS COMPONENTS, AND WHY THEY ARE CRITICAL FOR HUMANS AND FOR ALL LIFE ON EARTH

1

Christine Gibb, CBD and FAO

"Bio" means life and "diversity" means variety, so <u>biodiversity</u> (or biological diversity) is the incredible variety of living things in nature and how they interact with each other. It's one of the world's most precious treasures. Every human being, plant and animal contributes to the diversity, beauty and functioning of the Earth. This chapter introduces the concept and components of biodiversity, and some of the ways that biodiversity enriches our lives. The uses of biodiversity will be explored in later chapters.

> When you see text highlighted like this it indicates the word is in the glossary so you can find out more on what it means.

FROGS ON LILY PADS IN THE NATIONAL ARBORETUM, USA.
© Prerona Kundu (age 12)

CHAPTER 1 | What is biodiversity?

BIODIVERSITY, A THREE-PART CONCEPT

Biodiversity consists of all the many species of animals, plants, fungi, micro-organisms and other life forms and the variety that exists within each species.

It also includes the diversity present in ecosystems – or explained another way – the variation we see in the environment including landscapes, the vegetation and animals present in it, and the various ways in which these components interact with each other. Biodiversity is very complex and is often explained as the variety and variability of genes, species and ecosystems.

GENES

Genes are the units of heredity found in all cells. They contain special codes or instructions that give individuals different characteristics. Let's compare, for example, the genes coding for the necks of two different species: giraffes and humans. Even though both species have the same number of neck vertebrae (seven), the neck lengths of the two species are very different – approximately 2.4 metres for giraffes versus 13 centimetres for humans. This is because a giraffe's genes instruct each vertebra to grow up to 25 centimetres in length, whereas a human's instruct each vertebra to grow to less than two centimetres.

Genetic diversity occurs within a species and even within a variety of a given species. For instance, in a single variety of tomato, the genes of one individual may cause it to flower earlier than others, while the genes of another individual may cause it to produce redder tomatoes than other plants. Genetic diversity makes every individual unique. So in fact no two living things in nature are exactly the same. Chapter 3 delves into genetic diversity in greater detail.

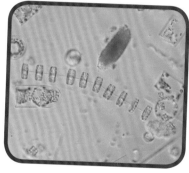

DIATOMS ARE MICROORGANISMS.
© C. Widdicombe, PML

YELLOW STRIPED LYCHNIS MOTH CATERPILLARS ON THE ROADSIDE IN AVEYRON, FRANCE.
© Clémence Bonnefous (age 8)

SPECIES

In our world you can find a dazzling array of animals, plants, fungi and micro-organisms. The different kinds of these are called 'species'. **A species is a group of similar organisms (individual living creatures such as spiders, walnut trees or humans) that can breed together and produce healthy, fertile offspring.** Although we may not think about it, we see various species as we go about our daily lives, such as humans, goats, trees and mosquitoes. Species diversity is the most obvious type of biodiversity. Our planet supports millions of species, many of which are not yet identified! At present, there are 310 129 known species of plants and 5 487 known species of mammals. There are perhaps millions of tiny organisms or micro-organisms that scientists have yet to identify. Chapter 4 explores species diversity, and answers species-related questions such as: why are species important?

SCHOOL OF FAIRY BASSLETS BETWEEN GIANT SEA FAN IN CORAL REEF.
© Korallenriff_139905

THE TEMPERATE RAINFOREST ON FRASER ISLAND, GREAT SANDY NATIONAL PARK, QUEENSLAND, AUSTRALIA IS A UNESCO WORLD NATURAL HERITAGE SITE.
© Michael Weber

KING PENGUINS IN THE SUBANTARCTIC REGION.
© Michael Weber

In the same way that humans live in communities, so too do animals, plants and even micro-organisms. **Where communities of plants and animals live together, and share their space, their land and their climate, they form an ecosystem.** Ecosystems are what many people call "the environment" or "nature". Chapter 5 provides an overview of ecosystems, and Chapters 6, 7, 8 and 9 take a closer look at biodiversity in several ecosystems. There are many kinds of ecosystems on Earth. Ecosystems can be small like puddles, or large like deserts, forests, wetlands, mountains, oceans, lakes and rivers.

ECOSYSTEMS

CHAPTER 1 | What is biodiversity?

BRINGING BIODIVERSITY TO THE WORLD STAGE

In 1992, an Earth Summit was held in Rio de Janeiro, Brazil, where governments, indigenous groups, women's groups, environmental groups, activists and other non-governmental organisations met to discuss the environment.

It was the largest international environmental meeting ever. In Rio, world leaders agreed that it was important to protect the environment for all people, including future generations. To reach this goal, the leaders decided to adopt three conventions (or agreements): the Convention on Biological Diversity (CBD), the United Nations Framework Convention on Climate Change (UNFCCC) and the United Nations Convention to Combat Desertification (UNCCD).

At the summit, participants agreed on the following definition of biodiversity:

"the variability among living organisms from all sources including, *inter alia*, terrestrial, marine and other aquatic ecosystems and the ecological complexes of which they are part; this includes diversity within species, between species and of ecosystems."

This is the official definition used by the Convention on Biological Diversity.

Convention on Biological Diversity

UNCCD

UNFCCC

IT'S ALL ABOUT THE
INTERACTIONS

Perhaps the most important characteristic of biodiversity is that all of the components are linked to each other.

For example, if a mouse eats a chemically-contaminated seed, it may survive, but if a hawk eats many mice that have eaten such seeds, the hawk may die from a lethal dose of the chemical. Because of their position in the food chain, top predators such as hawks are susceptible to such <u>biomagnification</u>, the accumulation of substances that increase in concentration up the <u>food chain</u>. Biodiversity linkages can also be beneficial: the restoration of coastal mangrove forest ecosystems provides an important nursery <u>habitat</u> for fish and other marine species, improves fisheries along the coastline, and protects human settlements from extreme weather events.

Similarly, the re-naturalisation of upstream rivers allows the recreation of a natural food chain, decreases the amount of mosquito larvae (thereby decreasing the incidence of malaria or similar mosquito-borne diseases), improves fisheries, and purifies water. If one level of biodiversity is interrupted, the other parts experience a ripple effect, which can be harmful or helpful to biodiversity.

The box: "Smaller Habitats Lead to Smaller Gene Pools" shows how the deterioration of an ecosystem negatively affects both species diversity and genetic diversity.

The box: "The Black Bear and the Salmon: Mighty Ecosystem Engineers" illustrates one positive example where two species play vital roles in engineering an ecosystem.

IMPROVING ONE ASPECT OF AN ECOSYSTEM BENEFITS THE WIDER ECOSYSTEM. A FORESTRY PROGRAMME TO REHABILITATE HILLSIDES IN NEPAL IMPROVED THE FLOW OF WATER FROM SPRINGS, WHICH, IN TURN IMPROVED CROP PRODUCTION.
© FAO/Giampiero Diana

CHAPTER 1 | What is biodiversity?

SMALLER HABITATS LEAD TO SMALLER GENE POOLS

The Florida Everglades in the USA is a unique ecosystem that was once home to many wading birds, mammals, reptiles, insects, grasses, trees and other species. It used to cover an area as large as England (over 9 300 square kilometres), but has shrunk over the years as more and more people moved there. The people also changed the ecosystem by building water management areas and canals, and filling in swampy areas for agriculture.

These ecosystem changes were bad for many species, including wood storks and Everglade kites.

The changes even affected the genes of some species such as the Florida panther! As suitable habitats were broken up into smaller and smaller pieces (scientists call this process "fragmentation"), only a few Florida panthers could survive.

With fewer breeding partners around, the variety in the gene pool (the total variety of genes available) declined. So the changes to the ecosystem negatively affected both species diversity and genetic diversity.

A KITTEN (TOP) AND ADULT (BOTTOM) FLORIDA PANTHER. THE KITTEN WAS MARKED WITH A TRANSPONDER CHIP, THE SAME KIND OF CHIP THAT IS USED TO IDENTIFY HOUSEHOLD PETS. THE ADULT WAS TAGGED WITH A RADIO COLLAR, WHICH HELPS BIOLOGISTS TRACK THE ANIMAL AND COLLECT DATA USED IN PANTHER CONSERVATION.
© Mark Lotz/Florida Fish and Wildlife Conservation Commission.

Sources: www.biodiversity911.org/biodiversity_basics/learnMore/BigPicture.html and www.nrdc.org/water/conservation/qever.asp

THE BLACK BEAR AND THE SALMON:
MIGHTY ECOSYSTEM ENGINEERS

Nutrients such as carbon, nitrogen and phosphorus generally flow downstream - from land to rivers then out to sea. But not always. In riparian forests (forests next to a body of water like a river, lake or marsh) in British Columbia, Canada, black bears help to transfer nutrients from the ocean back to the forest!

To understand how this nutrient transfer works, we need to know a little about the lifecycle of Pacific salmon. Pacific salmon are born in freshwater streams, where they feed and grow for several weeks. Once they're ready, they swim downstream and undergo physiological changes that allow them to survive in marine conditions.

The salmon spend up to several years in the ocean eating lots of crustaceans, fish and other marine animals (i.e. acquiring lots of nutrients from the ocean). Once they reach sexual maturity, the salmon leave the ocean and swim back to the exact freshwater stream where they were born. There, they spawn and die.

During the annual salmon run, black bears catch spawning salmon and carry them into the woods to eat. The nutrient transfer is significant. Each salmon offers two to 20 kg (sometimes even 50 kg) of essential nutrients and energy. One study in Gwaii Haanas, Canada found that each bear took 1 600 kg of salmon into the forest, eating about half. Scavengers and insects dined on the remains. The decaying salmon also released nutrients into the soil, feeding forest plants, trees and soil organisms.

In this way, vital nutrients are transferred from one ecosystem to another first by the salmon, then by the black bear.

Sources: ring.uvic.ca/99jan22/bears.html and www.sciencecases.org/salmon_forest/case.asp

THIS BLACK BEAR DINED ON A CHUM SALMON IT CAUGHT IN THE GEORGE BAY CREEK IN SKINCUTTLE INLET, HAIDA GWAII. LOOK CAREFULLY TO SPOT THE SALMON.
© Stef Olcen

CHAPTER 1 | What is biodiversity?

FIBRE AND CLOTHING
HARVESTING COTTON IN INDIA.
© Ray Witlin/World Bank

FOOD
SELLING VEGETABLES IN A MARKET IN KOREA.
© Curt Carnemark/World Bank

MEDICATION
MEDICINE FOR THE PRE-NATAL AND POST-NATAL CARE UNITS AT THE MAWANNALLA HOSPITAL IN SRI LANKA.
© Simone D. McCourtie/World Bank

THE BENEFITS OF

Biodiversity doesn't simply exist, it also has a function or purpose. <u>Ecosystems</u> provide things that humans benefit from and depend on. These things are called ecosystem <u>goods and services</u> and include the natural resources and processes that maintain the conditions for life on Earth. These ecosystem goods and services provide direct and indirect benefits, including the ones shown above. All life on Earth provides us with the food we eat, cleans the air we breathe, filters the water we drink, supplies the raw materials we use to construct our homes and businesses, is part of countless medicines and natural remedies, and many other things. Biodiversity helps to regulate water levels and helps to prevent flooding. It breaks down waste and recycles nutrients, which is very important for growing food. It protects us with "natural insurance" against future unknown conditions brought about by climate change or other events.

YOUTH AND UNITED NATIONS GLOBAL ALLIANCE

CULTURAL AND LEISURE BENEFITS
A CROSS-COUNTRY SKIER ENJOYS THE SPECTACULAR VIEWS IN THE ROCKY MOUNTAINS OF ALBERTA, CANADA.
© Christine Gibb

NUTRIENT CYCLING
RED WRIGGLER WORMS BREAK DOWN FRUIT, VEGETABLE AND PLANT SCRAPS, RETURNING NUTRIENTS TO THE SOIL.
© Christine Gibb

CLEAN AIR AND CLIMATE REGULATION
SMOKESTACKS FILL THE SKYLINE IN ESTONIA. TREES AND OTHER VEGETATION HELP FILTER OUT AIR-BORNE AND SOIL POLLUTANTS.
© Curt Carnemark/World Bank

LIVELIHOODS
FARMER WORKS ON THE TAMIL NADU IRRIGATION PROJECT IN INDIA.
© Michael Foley/World Bank

BIODIVERSITY

Many people also depend on biodiversity for their livelihoods and in many cultures natural landscapes are closely linked to spiritual values, religious beliefs and traditional teachings. Recreational activities are also enhanced by biodiversity. Think about when you go for a walk in the woods or along a river. Would it be as nice as if there was nothing but concrete buildings all around? Biodiversity is what enables ecosystems to continue to provide these benefits to people. As biodiversity is lost, we lose the benefits that ecosystems provide to people. This is why sustaining biodiversity is very intimately related to sustainable human development. Ecosystem services are further explained in Chapter 5 and Chapters 10 to 13 will further investigate the relationship between humans, biodiversity and sustainable development, and what different groups are doing to protect biodiversity.

THE YOUTH GUIDE TO BIODIVERSITY

CHAPTER 1 | What is biodiversity?

BIODIVERSITY IS ALL AROUND US

1 A LIVESTOCK HERDER PROTECTS HIS SHEEP IN INDIA.
© World Bank/Curt Carnemark

2 PLACES OF WORSHIP ARE OFTEN NESTLED IN BEAUTIFUL NATURAL AREAS THAT LEND THEMSELVES TO CONTEMPLATION AND PRAYER, SUCH AS THIS MONASTERY IN CHINA.
© Curt Carnemark/World Bank

3 TRADITIONAL SONGS AND DANCES TELL STORIES ABOUT THE LIFE HISTORIES AND CHARACTERS OF PLANTS AND ANIMALS, ESPECIALLY IN INDIGENOUS CULTURES. IN THIS PHOTO, DANCERS PERFORM AT A LOCAL CEREMONY IN BHUTAN.
© Curt Carnemark/World Bank

4 AN ECOTOURIST GUIDE INTRODUCES VISITORS TO NATURAL AND CULTURAL SITES IN UGANDA.
© FAO/Roberto Faidutti

5 BUTTERFLY HIKE IN BONN, GERMANY.
© Christine Gibb

6 TRADITIONAL FISHING IN MEXICO.
© Curt Carnemark/World Bank

7 BIOLOGIST PAULA KHAN WEIGHS A DESERT TORTOISE BEFORE RELEASING IT SOUTHEAST OF FORT IRWIN, CALIFORNIA, USA.
© Neal Snyder

8 CANOE TRIP IN FISH CREEK, USA.
© Christine Gibb

YOUTH AND UNITED NATIONS GLOBAL ALLIANCE

CONCLUSION

Biodiversity, the variety of life on Earth, is a vital ingredient of human survival and welfare. The importance of biodiversity goes beyond its value to human beings: all components of biodiversity have the right to exist. Unfortunately, all is not well for the planet's biodiversity. There are real threats to biodiversity, which will be explored in the next chapter.

LEARN MORE

:: Chapman. 2009. Number of Living Species in Australia and the World. Australian Government, Department of the Environment, Water, Heritage and the Arts: Canberra. www.environment.gov.au/biodiversity/abrs/publications/other/species-numbers/2009/pubs/nlsaw-2nd-complete.pdf

:: Resources on the Florida Everglades: www.biodiversity911.org/biodiversity_basics/learnMore/BigPicture.html and www.nrdc.org/water/conservation/qever.asp

:: Resources on ecosystem engineers: ring.uvic.ca/99jan22/bears.html and www.sciencecases.org/salmon_forest/case.asp

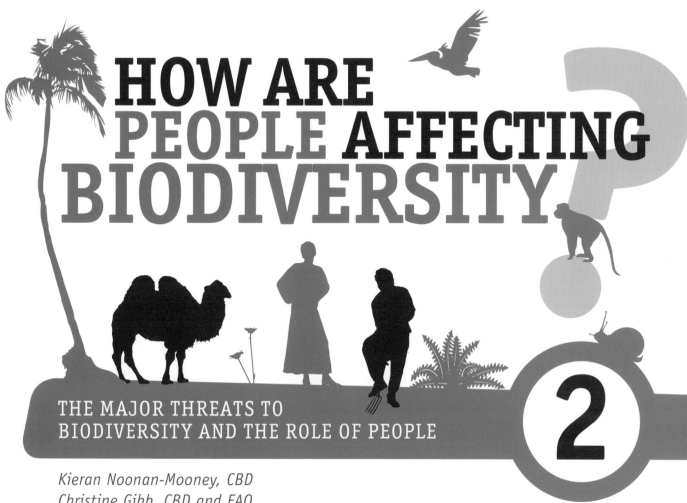

HOW ARE PEOPLE AFFECTING BIODIVERSITY?

THE MAJOR THREATS TO BIODIVERSITY AND THE ROLE OF PEOPLE

2

Kieran Noonan-Mooney, CBD
Christine Gibb, CBD and FAO

Every day we are faced with choices. As individuals we must decide what to eat, what to wear, how to get to school, and so on. Schools, businesses, governments and other groups also make choices. Some of these choices impact <u>biodiversity</u>, the variety of life on Earth. Sometimes our choices have positive impacts, for instance when we decide to use biodiversity sustainably or to protect it better. Increasingly, however, many of our actions are having negative consequences for biodiversity. In fact, human activities are the main cause of biodiversity loss.

LADYBIRD.
© Julia Kresse (age 15)

CHAPTER 2 | How are people affecting biodiversity?

The negative impacts of our actions have become so great that we are losing biodiversity more quickly now than at any other time in Earth's recent history. Scientists have assessed more than 47 000 **species** and found that 36 percent of these are threatened with **extinction**, the state whereby no live individuals of a species remain. In addition, extinction rates are estimated to be between 50 and 500 times higher than those observed from fossil records or the so-called "**background rate**". When species which are possibly extinct are included in these estimates, the current rate of species loss increases to between 100 and 1 000 times larger than the background rate!

The current rate of biodiversity loss has led many to suggest that the Earth is currently experiencing a sixth major extinction event, one greater than that which resulted in the extinction of the dinosaurs. However, unlike past extinction events, which were caused by natural disasters and planetary changes, this one is being driven by human actions.

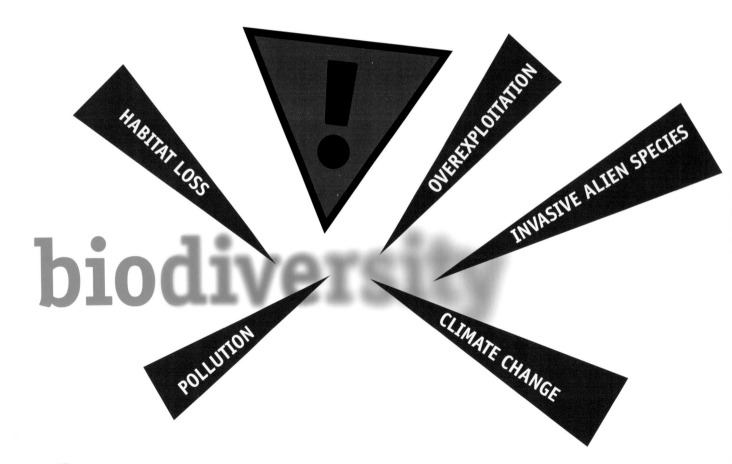

THE MAIN THREATS TO BIODIVERSITY

There are five main causes of biodiversity loss:

1. habitat loss
2. climate change
3. overexploitation
4. invasive alien species
5. pollution.

Each of these causes, or "direct drivers", puts tremendous pressure on biodiversity and often they occur at the same time in the same ecosystem or environment.

1 Habitat loss occurs when natural environments are transformed or modified to serve human needs. It is the most significant cause of biodiversity loss globally. Common types of habitat loss include cutting down forests for timber and opening up land for agriculture, draining wetlands to make way for new development projects, or damming rivers to make more water available for agriculture and cities. Habitat loss can also cause fragmentation, which occurs when parts of a habitat (the local environment in which an organism is usually found) become separated from one another because of changes in a landscape, such as the construction of roads. Fragmentation makes it difficult for species to move within a habitat, and poses a major challenge for species requiring large tracts of land such as the African forest elephants living in the Congo basin. Though some habitat loss is necessary to meet human needs, when natural habitats are changed or modified with little concern for biodiversity the results can be very negative.

POWER LINES CUT THROUGH THE BOREAL FOREST IN QUEBEC, CANADA. APPROXIMATELY 35 PERCENT OF QUEBEC'S BOREAL FOREST HAS BEEN AFFECTED BY HUMANS THROUGH HYDROELECTRIC POWER, FORESTRY, MINING, HUNTING, FISHING AND RECREATIONAL ACTIVITIES.
© Allen McInnis/Boreal Communications

CHAPTER 2 | How are people affecting biodiversity?

2 <u>Climate change</u>, which is caused by a build-up of greenhouse gases such as carbon dioxide in the Earth's atmosphere, is a growing threat to biodiversity. Climate change alters the climate patterns and ecosystems in which species have evolved and on which they depend. By changing the temperature and rain patterns species have become accustomed to, climate change is changing the traditional ranges of species. This forces species to either move in order to find favourable conditions in which to live, or to adapt to their new climate. While some species may be able to keep up with the changes created by climate change, others will be unable to do so. Biodiversity in the polar regions (see box: "Arctic Sea Ice and Biodiversity") and mountain ranges is especially vulnerable to climate change.

3 <u>Overexploitation</u>, or <u>unsustainable use</u>, happens when biodiversity is removed faster than it can be replenished and, over the long term, can result in the extinction of species. For example:
- the once-plentiful cod fishery off the coast of Newfoundland, Canada has all but disappeared because of overfishing;
- freshwater snakes in Cambodia are declining from hunting pressure;
- *Encephalartos brevifoliolatus*, a cycad, is now extinct in the wild after being overharvested for use in horticulture;
- Overexploitation, especially when combined with destructive harvesting practices, is a major cause of biodiversity loss in certain ecosystems.

4 ▶ Invasive alien species (IAS), or species that have spread outside of their natural habitat and threaten biodiversity in their new area, are a major cause of biodiversity loss. These species are harmful to native biodiversity in a number of ways, for example as predators, parasites, vectors (or carriers) of disease or direct competitors for habitat and food.

In many cases invasive alien species do not have any predators in their new environment, so their population size is often not controlled (see box: "The Troublesome Toad"). Some IAS thrive in degraded systems and can thus work in conjunction with or augment other environmental stressors. IAS may also cause economic or environmental damage, or adversely affect human health.

The introduction of invasive alien species can be either intentional, as with the introduction of new crop or livestock species, or accidental such as when species are introduced through ballast water or by stowing away in cargo containers. Some of the main vectors (carriers) for IAS are trade, transport, travel or tourism, which have all increased hugely in recent years.

FELLING OF TREES IS A MAJOR ECOSYSTEM DISTURBANCE AFFECTING NUMEROUS SPECIES.
© FAO/L.Taylor

ANIMALS ARE VARIOUSLY AFFECTED BY DIFFERENT TYPES OF POLLUTION. OIL SPILLS CAN DEVASTATE FISH, SEA TURTLE AND MARINE BIRD POPULATIONS.

BROWN PELICANS CAPTURED AT GRAND ISLE, LOUISIANA, USA, WAIT TO BE CLEANED.
© International Bird Rescue Research Centre

SOME ANIMALS, SUCH AS THIS HERMIT CRAB IN THAILAND, MAKE THE BEST OF A BAD SITUATION AND USE TRASH ON THE BEACH AS A MAKESHIFT HOME.
© Alex Marttunen (age 12)

THE YOUTH GUIDE TO BIODIVERSITY

CHAPTER 2 | How are people affecting biodiversity?

5 ▶ The final driver of biodiversity loss is <u>pollution</u>. Pollution, in particular from nutrients, such as nitrogen and phosphorus, is a growing threat on both land and in aquatic ecosystems. While the large-scale use of fertilisers has allowed for the increased production of food, it has also caused severe environmental damage, such as <u>eutrophication</u> (see box below).

EUTROPHICATION

In eutrophic water bodies, such as lakes and ponds, the concentration of chemical nutrients is so high that algae and plankton begin to grow rapidly. As these plants grow and decay, the water quality and the amount of oxygen in the water decline. These conditions make it difficult for many species to survive. The excess nutrients that cause this situation mostly come from fertilisers, erosion of soil containing nutrients, sewage, atmospheric nitrogen deposition and other sources.

EUTROPHICATION IS CAUSED BY TOO MANY NUTRIENTS IN THE WATER.
© F. Lamiot

YOUTH AND UNITED NATIONS GLOBAL ALLIANCE

THE TROUBLESOME TOAD
Saadia Iqbal, Youthink!

It all started with some beetles that were destroying sugarcane crops in Australia. A type of toad called the cane toad was brought in from Hawaii, with the hope that they would eat the beetles and solve the problem. Well, the toads left the beetles alone, but ate practically everything else, becoming full-fledged pests in their own right.

Now they are running amok, preying on small animals and poisoning larger ones that dare to try eating them. Scientists are still trying to figure out what to do.

Source: australianmuseum.net.au/Cane-Toad

CANE TOAD.
© H. Ehmann/Australian Museum

CHAPTER 2 | How are people affecting biodiversity?

ARCTIC SEA ICE AND BIODIVERSITY

In the Arctic, ice is the platform for life. Many groups of species are adapted to life on top of or under ice. Many animals use sea ice as a refuge from predators or as a platform for hunting. Ringed seals need certain ice conditions in the spring for reproduction, while polar bears travel and hunt on the ice. Algae even grow on the underside of ice floating on the ocean. Ice is also the surface for transportation and is the foundation of the cultural heritage of the native Inuit people.

The pattern of annual thawing and refreezing of sea ice in the Arctic Ocean has changed dramatically in the first years of the twenty-first century. The extent of floating sea ice, measured every September, has declined steadily since 1980 (shown by the red trend line). Not only is the ice shrinking, but it is also much thinner.

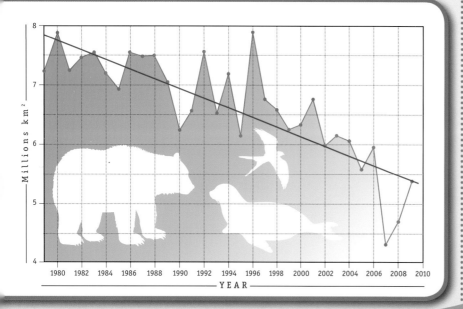

Adapted from: Global Biodiversity Outlook 3, 2010

YOUTH AND UNITED NATIONS GLOBAL ALLIANCE

CONCLUSION

Though biodiversity loss is occurring at a rapid rate, examples from all over the world show that people are beginning to make choices and take actions that benefit biodiversity. However, we need more action if further biodiversity loss is to be averted. It's important to carefully consider the choices you make and their impacts, and to encourage other groups such as businesses and governments to do the same. The rest of this guide will help inform you about issues to consider, steps you can take and examples of positive action for biodiversity.

LEARN MORE

- Global Biodiversity Outlook 3: gbo3.cbd.int
- Global Environment Outlook 4: www.unep.org/geo/geo4/media
- Invasive alien species:
  www.cbd.int/iyb/doc/prints/factsheets/iyb-cbd-factsheet-ias-en.pdf ,
  www.cbd.int/doc/bioday/2009/idb-2009-childrens-booklet-en.pdf and
  australianmuseum.net.au/Cane-Toad
- Climate change:
  www.cbd.int/doc/bioday/2007/ibd-2007-booklet-01-en.pdf

GENES & GENETIC DIVERSITY

EXPLAINING GENETIC DIVERSITY AND HOW IT CONTRIBUTES TO OUR LIVES AND FUTURES

3

Cary Fowler, Charlotte Lusty and Maria Vinje Dodson, Global Crop Diversity Trust

Every cell of every individual of every species contains genes. No two individuals have exactly the same genes – that is, unless they are clones. The discovery of genes, what they look like and how they are passed from parent to child, led to a revolution in science in the nineteenth and twentieth centuries.

DIFFERENT VARIETIES OF MAIZE.
© FAO

CHAPTER 3 | Genes and genetic diversity

GENES AND DIVERSITY

It all started with a scientist and monk called Gregor Mendel. In the mid 1800s Mendel experimented on peas in the gardens of the Abbey of St. Thomas in Brno, now in the Czech Republic.

In one experiment, Gregor Mendel selected a tall plant, then crossed (or bred) it with a short plant. He observed how high the plant offspring grew and then how high the next generation of offspring grew. Looking at the pattern of tall-to-short plants, he was able to describe the basic laws of genetics. These laws, roughly summarised, state that when two parents reproduce each will pass on only half of their genetic material to their offspring. Each offspring will have exactly the same number of genes from their mother, and their father. These genes are passed on quite randomly, so all offspring inherit something different and no two siblings are quite the same, as can be seen in the figure on the opposite page. There are exceptions, however. Identical twins are nearly genetically identical, thanks to a rare event in nature where one fertilised egg divides and develops into two offspring. In certain plant families, clones are actually the norm. For instance, when a new bamboo or banana plant shoots up from the side of a parent plant, it is a clone.

OPPOSITE PAGE: FLOWER OF PEA PLANT.
© Giulia Tiddens

A — A tall pea plant (TT) is crossed with a short pea plant (tt).

B — Offspring get one gene for a given trait, like tallness, from each parent. The different traits do not blend. The result of this cross is two tall plants.

C — When these two tall plants are crossed, however..

D — ...the result is three tall plants and one short plant because tallness is a dominant trait and shortness is recessive. But recessive traits can crop up in later generations if two copies of a recessive gene are present.

CHAPTER 3 | Genes and genetic diversity

GENETIC DIVERSITY

Putting it simply, genes produce <u>traits</u>. A trait is a characteristic such as curly hair, freckles or blood type. Genes act singularly or in combination to produce traits, many of which are vevry obvious. Human eye colour, for instance, is determined by a combination of genes, giving colours of blue, green, brown, hazel, grey, chestnut and variations inbetween. This is <u>genetic diversity</u>.

When you look at the eyes of the population of a town or country in Europe or North America, there may be great variation in eye colour. Elsewhere, for instance in parts of Africa and Asia, eye colour may not vary much at all.

GENETIC DIVERSITY WITHIN A SPECIES – HUMAN EYES.
(From left to right)
© [farm2.iistatic.flickr.com/1320/1129245762_f616924190.jpg]
© [farm4.static.flickr.com/3097/2773264240_f91e272799.jpg]
© Cristina Chirtes [www.flickr.com/photos/p0psicle/2463416317]
© Cass Chin [www.flickr.com/photos/casschin/3663388197]

The eye colour example shows diversity within a single species – the human species. But there's also genetic diversity within a group of species – pigeons, for instance! If you travel to an island or a forest in the tropics, you will almost certainly find pigeons. You will recognise that they are pigeons, but they are not the same as the pigeons you know from back home. They are genetically quite different, although clearly related. There are many more diverse groups of species than pigeons, especially in the insect world!

TOMATOES: THE SAME SPECIES BUT A LARGE VARIETY OF
SHAPES, COLOURS, SMELLS AND ESPECIALLY TASTES.
© Reuben Sessa

THERE ARE 4 810 DIFFERENT SPECIES OF FROGS
WORLDWIDE. HERE ARE FOUR EXAMPLES.
(From left to right)
© [farm1.static.flickr.com/89/232636845_5ca3c4fe51.jpg]
© [farm3.static.flickr.com/2761/4330810650_47ed959dfd.jpg]
© [farm2.static.flickr.com/1405/1395010192_e3f85c9c7c.jpg]
© Diego Adrados (age 14)

In a similar way, we can measure the total genetic diversity of an **ecosystem**. Some ecosystems are more diverse than others. A small area of forest on the Atlantic coast of Brazil contains more plant and animal species, and therefore more genetic diversity, than the entire USA.

CHAPTER 3 | Genes and genetic diversity

FORCES OF EVOLUTION

There hasn't always been as much genetic diversity on Earth as there is today. It evolved from practically nothing. There are four ingredients to evolution: natural selection, variation, inheritability and time. The combination of these ingredients accounts for the evolution of species – everything from whales to drug-resistant bacteria!

Let's go back to eye colour, only let's imagine there is a forest that is home to a population of pigeons with different eye colours – let's say some have green eyes and others grey eyes. By chance, the green-eyed pigeons are particularly good at seeing in the dark. In this imaginary forest, the pigeons feed all day on delicious figs only available on a small number of fig trees. During the night, the birds rest. But a change in the climate drives a new species of pigeon-hunting eagle into the area. The eagle's talent for diving from a great height onto pigeons feeding in the fruit trees depends on good daytime visibility.

The pigeons soon begin to hide in the daytime and feed at night. The grey-eyed pigeons have trouble finding fruit at night and are caught by the eagle during the day. The pigeons with green eyes, however, are able to eat without the threat of the eagle because they can feed at night. The green-eyed pigeons' green-eyed offspring are more successful! They live longer and have more offspring. As the grey-eyed pigeons disappear, the pigeon population starts to change. Over time, most of the newborn pigeons have the same eye colour – green!

This story illustrates that where there is *variation* in a population (e.g. in eye colour), new or existing pressures (e.g. eagles) *select* specific *inheritable traits* passed on from one generation to the next (e.g. green eyes) that convey an advantage. Over *time* (e.g. several generations of pigeons) the population changes and the species *evolves*. Where there are no pigeon-hunting eagles, the pigeons can still thrive whatever their eye colour. Eventually after much time has passed, the green-eyed pigeons, with their skills at feeding in the night and avoiding eagles, could become a completely independent species.

YOUTH AND UNITED NATIONS GLOBAL ALLIANCE

WHO'S MORE DIVERSE, HUMANS OR MAIZE?

Although humans may look diverse they are actually genetically very similar. In fact there is more diversity in a single field of maize than there is in the entire human population! Can you imagine the potential for maize to develop into many different types of crops?

MAIZE.
© Global Crop Diversity Trust

© Romain Guy/www.flickr.com/230420772

© Richard [www.flickr.com/photos/68137880@N00/3082497758]

© FAO/Riccardo Gangale

© Reuben Sessa

THE YOUTH GUIDE TO BIODIVERSITY

CHAPTER 3 | Genes and genetic diversity

GENETIC DIVERSITY IN USE

Diversity plays a key role in evolution and the survival of species. There are endless examples, from the dinosaur to the dodo, where a species lacked variation or the ability to adapt to pressures in its environment and became extinct. In the case of humans, we have so successfully adapted to our environment that we are now one of the major factors shaping and changing the planet. Despite this fact, we are still highly dependent on diversity at all levels.

Firstly, at an ecosystem level, diversity provides our habitats and environments. At the most fundamental level, plants provide the oxygen in the atmosphere. Our diverse habitats provide soils, water, homes, cover from the sun or wind, and many other services to support life.

Secondly, species diversity is important because humans eat an omnivorous diet and live in diverse environments around the world. Unlike cows and pandas, if we always eat the same, single type of food, we become sick because of a lack of essential nutrients in our diet. Diversity in our food systems is important to keep us alive. Diversity has enabled humans to colonise and thrive in many diverse living conditions around the world. Diversity also provides our medicines, timber, paper, fuel, raw products for manufacturing and just about everything else upon which human civilisation is based.

Genetic diversity and the variation of traits that it provides allow individual species to adapt to changes in the environment. All species, such as humans, are under the constant threat of a new flu or other disease, as well as weather and temperature changes. Food may turn up reliably in supermarkets and shops but behind the scenes scientists and farmers are working constantly to keep up yields to meet the demand. Genetic diversity is the basis on which crops adapt and evolve in the face of challenges.

Over the past 12 000 years, farmers have selected individual plants that yield more, taste better, or survive well under pressure. In different places or times, thousands of farmers used the seed from their preferred plants to sow the next season's crop. In this way people have shaped crops to meet their needs under specific conditions. Thousands of crop varieties have been developed around the world. Modern-day breeders similarly select plants with specific traits, using various techniques or tools to speed up the process and to produce the high-yielding varieties that we are likely to buy in supermarkets. The box "The Making of Our Daily Bread" illustrates the importance of genetic diversity to wheat farmers.

THE MAKING OF OUR **DAILY BREAD**

In the 1940s, farmers were losing much of their wheat crop to a fungus known as stem rust. The fungal spores are carried by the wind from field to field. The spores can land on any part of the wheat plant and infect it, forming pustules on the stem and leaves, and causing the plant to produce much less seed, if not to die altogether.

Breeders screened wheat **genebanks** – storehouses of genetic diversity – for plants that do not appear to get the symptoms of the disease when grown in the presence of the fungal spores. They crossed popular wheat varieties that die from stem rust with the **wild relatives** that seem to resist the disease to breed new forms of disease-resistant wheat that produce good yields. The new varieties were taken up enthusiastically by farmers, and spread worldwide. One of the scientists responsible, Norman Borlaug, went on to win a Nobel Prize for his efforts. There is probably not one single reader of this book who has not benefited from Norman Borlaug's work.

In 1999, a new stem rust appeared in Uganda and spread into the Middle East. Plant breeders are screening all their genetic resources once more to find wheat varieties with resistance to the new disease.

It is important to realise that a new devastating disease in a major crop is not news – this is business as usual in the world of agriculture! So it's important to keep a large genetic diversity to be able to produce new varieties which are resistant to these new diseases.

WHEAT GROWING IN SPAIN.
© Bernat Casero

CHAPTER 3 | Genes and genetic diversity

GENETIC DIVERSITY UNDER THREAT

A place without genetic diversity is chronically fragile and poised for disaster. In 1845, a deadly disease destroyed the potato crop, the main crop of the rural poor in Ireland, leading to the starvation or migration of two million people. There have been many events like this in human history.

Apart from dramatic famines or **extinctions**, there is the more gradual threat of **genetic erosion** or loss of genes and the traits that they produce. Present-day agricultural, forestry or aquaculture systems are more **homogeneous** (similar) over wide geographical areas than ever before, cultivating a smaller number of the same species and varieties. But diversity is still very much appreciated and used, especially in areas where people depend entirely on their crops throughout the year as a source of food. How else does a family eat if one crop fails? A very rich source of crop diversity is also hidden in remaining wild habitats, where the **wild relatives** of crop species can still be found.

THE FLOWERS OF CULTIVATED POTATOES (**left**) AND WILD POTATOES (**right**).
© Martin LaBar
© Arthur Chapman

In the volatile world in which we live, diversity is an important factor. A large number of scientists, breeders and farmers are working to safeguard biodiversity to allow humans to meet the challenges of an unpredictable future. This can be achieved in different ways.

GENETIC DIVERSITY
FOREVER

Conserving seeds is more complicated than you might imagine. The easy option involves packing materials into airtight containers and storing them at low temperatures. For plants that don't produce seeds, the materials are conserved as miniature plantlets in glass tubes in the laboratory or as tissue samples frozen to ultra-low temperatures in liquid nitrogen. In both cases, the materials may remain dormant for several decades although they still need to be checked regularly to ensure that they are not deteriorating.

A genebank is not like a library where browsers can come and read the books or ask for particular titles or authors. You cannot tell from a seed how the plant will grow, whether it will deal with diseases or certain climates, nor how the harvested crop will taste. One of the most important aspects of genebanking is to test the plants and document meticulously their traits and characteristics. Genebanks keep thousands of samples. For instance, there are more than 250 000 entries for maize in different genebanks around the world. That is a lot of seeds to look through to find the maize that may work for you!

The ultimate safe place for crop diversity is the Svalbard Global Seed Vault in Norway. Here, within the Arctic Circle dug into a frozen mountainside, safe from hurricanes, floods, electricity blackouts and wars, the genebanks of the world are depositing samples from their collections as a safety backup. So far, more than 500 000 seed samples are in storage. Whether these seeds will ever be needed is anyone's guess. This diversity represents a multitude of options that we can provide to the people of the distant or maybe not-so-distant future.

THE SVALBARD SEED VAULT CUT WITHIN THE MOUNTAIN IN NORWAY.
© Global Crop Diversity Trust
ENTRANCE TO THE SVALBARD SEED VAULT IN NORWAY.
© Svalbard Global Seed Vault/Mari Tefre

Communities or individual farmers have been safeguarding their seeds and animal breeds since organised agriculture began. With a better knowledge of genetics and evolution, scientists began to recognise the significance of diversity and collect rare breeds and crop varieties. Today there is a worldwide network of institutes that conserves the seeds, roots and tubers of crops in genebanks.

CHAPTER 3 | Genes and genetic diversity

The International Maize and Wheat Improvement Center **(CIMMYT)** is a research and training center with the objective to sustainably increase the productivity of maize and wheat systems to ensure global food security and reduce poverty.

The Tropical Agricultural Research and Higher Education Center **(CATIE)** works on increasing human well-being and reducing rural poverty through education, research and technical cooperation, promoting sustainable agriculture and natural resource management. One of its areas of work is improving coffee production systems.

THE WORLD'S LARGEST *EX SITU* COLLECTIONS

This map shows a selection of the important international crop collections.

Source: Global Crop Diversity Trust.

The International Center for Tropical Agriculture **(CIAT)** has the world's largest genetic holdings of beans (over 35 000 materials), cassava (over 6 000), and tropical forages (over 21 000), obtained or collected from over 141 countries.

The International Potato Center **(CIP)** is a root and tuber research-for-development institution delivering sustainable solutions for the pressing global problem of hunger. It has a large gene bank of potato varieties and other root species.

The Brazilian Agricultural Research Corporation **(EMBRAPA)** develops technologies and identifies practices to improve agricultural production. It is a major research institution for cassavas as well as other tropical fruits such as pineapples, acerolas, bananas, citrus fruit, papayas, mangos and passion fruit.

Bioversity International uses agricultural biodiversity to improve people's lives by researching solutions for three key challenges: sustainable agriculture, nutrition and conservation. Bioversity maintains the international Germplasm Collection of both improved varieties and wild species of bananas.

The Vavilov Research Institute of Plant Industry **(VIR)** undertakes research and development for numerous types of crops and varieties. It supports the collection and maintenance of gene banks for numerous crop species, including barley.

The World Vegetable Center **(AVRDC)** works in four main areas: germplasm, breeding, production and consumption. AVRDC maintains the world's largest public vegetable genebank with more than 59 294 entries from 155 countries, including about 12 000 of indigenous vegetables.

The International Institute of Tropical Agriculture **(IITA)** works on enhancing crop quality and productivity. It works on assessing the growing properties and nutritional content of new varieties of yam and works with other crops such as cowpeas, soybeans, bananas/plantains, cassavas and maize.

The International Center for Agricultural Research in the Dry Areas **(ICARDA)** contributes to the improvement of crops such as chickpea, bread and durum wheats, kabuli, pasture and forage legumes, barley, lentil and faba bean. Other work includes supporting improvement of on-farm water-use efficiency, rangeland and small-ruminant production.

The International Crops Research Institute for the Semi-Arid Tropics **(ICRISAT)** conducts agricultural research for development in Asia and sub-Saharan Africa. Its genebank serves as a world repository for the collection of germplasm of a number of crops including sorghum.

The International Rice Research Institute **(IRRI)** develops new rice varieties and rice crop management techniques that help rice farmers improve the yield and quality of their rice in an environmentally sustainable way. IRRI maintains the biggest collection of rice genetic diversity in the world, with more than 113 000 types of rice, including modern and traditional varieties, as well as wild relatives of rice.

THE YOUTH GUIDE TO BIODIVERSITY

CHAPTER 3 | Genes and genetic diversity

WHAT CAN YOU DO TO CONSERVE GENETIC DIVERSITY?

- Visit a local Farmers' Market. The farmers usually grow and sell local varieties of fruits and vegetables that you won't find in the supermarket. By buying their products you are encouraging the farmers to continue growing genetically distinct varieties.

- Try growing local fruits and vegetables at home. If you grow two different types of tomatoes, you may see how they sprout, flower and fruit at different times. You will also see how they taste different too!

- You can grow plants native to your area in your garden.

- You can encourage your school or neighbourhood to set up community gardens.

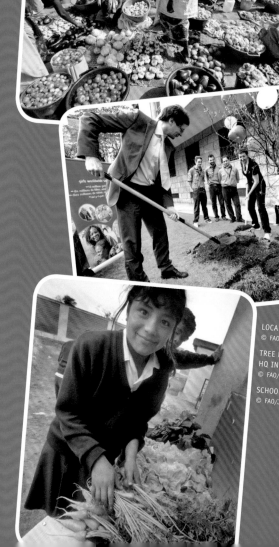

LOCAL MARKET IN GAMBIA.
© FAO/Seyllou Diallo

TREE PLANTING CEREMONY AT FAO HQ IN ITALY.
© FAO/Alessandra Benedetti

SCHOOL GARDEN IN PERU.
© FAO/Jamie Razuri

:: Turn vacant or abandoned areas into green lots, where everyone can plant fresh fruits and vegetables, flowers and any other plants they would like to grow. For inspiration visit www.nybg.org/green_up.

:: You can join a group that helps to conserve diversity or to protect the environment. For example, look for groups that plant trees, take care of animals or run urban farms or gardens.

:: Reducing waste, recycling garbage and using environmentally-friendly cleaning products all help to protect the environment and reduce the threat to endangered species.

:: Give a presentation about genetic diversity at your school. You can, for example, talk about the Svalbard Global Seed Vault, where hundreds of thousands of food seeds are being kept safe for the future. You can read more about it at www.croptrust.org.

COMMUNITY SCHOOL GARDEN IN HAITI.
© FAO/Thony Belizaire

SCHOOL ORCHARD IN INDIA.
© FAO/Jon Spaull

STAFF CAMPAIGN TO ENCOURAGE GREEN HABITS AT FAO IN ITALY.
© FAO/Giulio Napolitano

GROUP WORK IN THE CENTRAL AFRICAN REPUBLIC.
© FAO/Riccardo Gangale

SPECIES: THE CORNERSTONE OF BIODIVERSITY

AN EXAMINATION OF HOW SPECIES DIVERSITY IS THE KEY TO A HEALTHY PLANET, AND A CLOSER LOOK AT A MAJOR TOOL USED IN BIODIVERSITY CONSERVATION

Kathryn Pintus, IUCN

So far, we've had a look at genetic diversity, and we've learned that genes are responsible for the wide variety of species that exist on Earth. But what exactly is a species?

A species is a basic biological unit, describing organisms which are able to breed together and produce fertile offspring (offspring that are able to produce young). The above statement is a fairly widely accepted definition, and in some cases it is easy enough to determine whether two organisms are separate species simply by looking at them; the mighty blue whale is clearly not the same species as the fly agaric mushroom.

A BANANA SLUG EATING A RASPBERRY AT A CAMPSITE IN REDWOOD NATIONAL PARK, CALIFORNIA, USA.
© Anthony Avellano (age 14)

CHAPTER 4 | Species: the cornerstone of biodiversity

However, the situation is not always quite so straightforward. The science of describing and classifying organisms is called <u>taxonomy</u>, and this provides us with a common language that we can all use to communicate about species, but it can get rather complicated! Biology is split into several fields, including botany, zoology, ecology, genetics and behavioural science, and scientists from each of these branches of biology will have slightly differing definitions for what constitutes a species, depending on the focus of their specialty. For instance, some definitions will be based on <u>morphology</u> (what it looks like), others on <u>ecology</u> (how and where it lives), and others still on <u>phylogenetics</u> (using molecular genetics to look at evolutionary relatedness). For this reason, when considering two <u>organisms</u> which on the surface may look almost identical, scientists sometimes disagree as to how to classify them. Are they individuals of the same species or are they two completely separate species? Or are they perhaps <u>subspecies</u>? Having said that, taxonomy can be very useful (see the box: "How Does Taxonomy Help Biodiversity?" for more details).

BLUE WHALE.
© bigsurcalifornia

To complicate matters further, some individuals of the *same* species may look considerably different from one another, perhaps due to their sex or as a result of their geographical distribution. The trait whereby males and females look different from one another is known as <u>sexual dimorphism</u>, and can be seen in many species, particularly in birds.

Scientific disagreements aside, there are about 1.78 million described species on Earth, with millions more out there that we don't even know about yet. That's an incredible amount of biodiversity, but unfortunately much of it is being lost, and it is possible that we are losing some species before we even have the chance to discover them.

FLY AGARIC MUSHROOMS AT LOCKERBROOK FARM IN DERBYSHIRE, UK.
© Roger Butterfield

40 YOUTH AND UNITED NATIONS GLOBAL ALLIANCE

HOW DOES TAXONOMY HELP BIODIVERSITY?

Junko Shimura, CBD

Biodiversity, or life on Earth, is disappearing at an unprecedented rate as a result of human activities. Decisions must be taken now to reverse this trend. But how do decision-makers decide where to establish protected areas, places that receive special protection because of their environmental or cultural value, if they don't know what needs protecting? How can regulators identify and combat harmful invasive alien species if they cannot distinguish them from native species? How can countries use their biodiversity if they don't know what biodiversity exists within their borders?

The field of taxonomy answers these questions and more!

Taxonomy is the science of naming, describing and classifying organisms and includes all plants, animals and micro-organisms of the world. Using morphological, behavioural, genetic and biochemical observations, taxonomists uncover evolutionary processes and study relationships among species.

Unfortunately, taxonomic knowledge is far from complete. In the past 250 years of research, taxonomists have named about 1.78 million species of animals, plants, fungi and micro-organisms. Though the total number of species is unknown, it is probably between five and 30 million, which means only six to 35 percent of the Earth's species have been scientifically identified. Without thorough taxonomic knowledge, it's very difficult to have effective conservation and management of biodiversity.

Governments, through the Convention on Biological Diversity, noted this "taxonomic impediment" to the sound management of biodiversity. In 1998, they launched the Global Taxonomic Initiative (GTI) to fill the knowledge gaps in our taxonomic system, to fix the shortage of trained taxonomists and curators, and to address the impact these deficiencies have on our ability to conserve, use and share the benefits of biodiversity.

For more information about scientific naming see Annex B.

CHAPTER 4 | Species: the cornerstone of biodiversity

THE IMPORTANCE OF SPECIES

Now that you have a better idea of what a species is, you might be asking yourself the following question: why are species important?

Species play a vital role as building blocks of biodiversity, interacting to form the ecosystems upon which we all depend for survival, and providing us with what are known as ecosystem goods and services (discussed extensively in the next chapter).

Goods are things that we can physically use or sell, including food, fuel, clothes and medicine, whilst services include the purification of water and air, crop pollination and cultural values.

HEN AND CHICKS FROM NICARAGUA.
© FAO/Saul Palma

GOAT HERDING IN LEBANON.
© FAO/Kai Wiedenhoefer

SPECIES ARE A SOURCE OF FOOD, FOR EXAMPLE THESE GROUND NUTS (PEANUTS) GROWN IN CHINA.
© FAO/Florita Botts

THE DOMESTIC COW AND ITS ANCIENT RELATIVE, THE AUROCH. THE LAST AUROCH DIED IN 1627 IN THE JAKTORÓW FOREST, POLAND; PREHISTORIC CAVE PAINTINGS, SUCH AS THOSE IN THE LASCAUX CAVES OF FRANCE, ARE THE ONLY IMAGES OF AUROCHS THAT EXIST TODAY.
© Prof Saxx/Wikimedia Commons
© FAO/Giuseppe Bizzarri

SPECIES BIODIVERSITY AND GOODS

Many of the goods we obtain come from domesticated species, including cows, pigs and sheep, and various agricultural crops such as wheat, rice and corn. All of these domesticated species originally descended from wild ones, which were selected and bred for specific purposes. The food produced from domesticated species sustains the lives of billions of people around the world, by forming the essential components of our daily diets. Despite there being thousands of species that we could potentially eat, we routinely consume only a small handful of these!

Wild species are just as important as domesticated species; people across the globe rely on marine, freshwater and terrestrial <u>ecosystems</u> for food and materials they need to survive. The oceans, for instance, cover more than 70 percent of the Earth's surface, and house an astonishing array of biodiversity, some of which provides essential food and income for millions of people. Freshwater ecosystems are just as valuable to humans; an estimated 126 000 described species including fish, molluscs, reptiles, insects and plants rely on freshwater habitats, many of which are an extremely important component of the <u>livelihoods</u> of local people. Rainforests such as those found in South America contain thousands upon thousands of species, some of which are extremely important in both modern and traditional medicine. As such, healthy biodiversity is essential to human well-being.

DID YOU KNOW? Since agriculture began about 12 000 years ago, roughly 7 000 plant species have been used for human consumption.

DID YOU KNOW? More than 70 000 different plant species are used in traditional and modern medicine.

CHAPTER 4 | Species: the cornerstone of biodiversity

SPECIES BIODIVERSITY AND SERVICES

Services provided by species include water purification, which is carried out by molluscs such as clams and mussels. These molluscs are very common in river systems, and purify the water by filtration, making it safer to drink. Dragonflies also play an important role in freshwater ecosystems, by acting as indicators of water quality. If pollution becomes a problem in an area, dragonflies will be the first to be affected, and so a reduction in their numbers could indicate a reduction in water quality. This early warning mechanism could prove vital, as action can then be taken to resolve the problem before other species (including humans) become affected.

Wild species also provide valuable services such as pollination, the process of transferring pollen to enable reproduction in plants. Pollination is needed by most of the world's land-based plant life. The "Animal Pollination" box discusses some of the adaptations various plant species have evolved to attract different species of animal **pollinators**.

DID YOU KNOW?
The global value of wetlands is US$ 17 trillion!

DRAGONFLY.
© raymondbPhotos

ANIMAL POLLINATION
Nadine Azzu, FAO

Pollination is a very important ecosystem service, without which many of the plants that we use for food could not grow. Pollination can occur in three main ways: self-pollination, wind pollination and pollination by animals. But let's talk about pollination by *animals*. There are many types of animal pollinators, including: insects (e.g. bees, wasps, flies, beetles, moths and butterflies), birds (e.g. hummingbirds), and mammals (e.g. bats and the Australian honey possum). Insects are the most common pollinator, especially because they are small and can easily fly from flower to flower.

Pollination by animals is a very particular and wondrous event. For a plant to be pollinated, the habits and physical characteristics of the animal (e.g. its mouth shape, its ability to see and smell, and even the way it moves around) must be well-matched with the habits and physical characteristics of the flower (e.g. colour, scent and structure). The need for a "perfect match" is one of the reasons why different types of pollinators pollinate different plants. For example:

- Bees are attracted by a flower's colour, scent and especially nectar (the bees' food).

- Beetles, who do not see well, are generally attracted to flowers with a strong scent.

- Butterflies, who pollinate during the day, rely mainly on the visual stimulus provided by the plant (in other words, colour).

- Moths tend to pollinate at night and may rely less on visual cues than on olfactory cues (smell). To catch the attention of their moth pollinators, some plant species emit varying scent intensities throughout the day, with a stronger scent during the evening when moths are active. An example of such a flower is the night blooming jasmine.

NIGHT BLOOMING JASMINE PRODUCES A STRONGER SCENT AT NIGHT TO ATTRACT MOTHS.
© Asit K. Ghosh/Wikimedia Commons

CHAPTER 4 | Species: the cornerstone of biodiversity

Some red flowers do not have a strong scent at all – for these flowers, hummingbirds are an ideal pollinator. Why is that? Because the vision of hummingbirds is particularly good at seeing red in the colour spectrum, and they also have an underdeveloped sense of smell so, a flower does not need to have such a strong scent for hummingbirds to find it.

Other plant species have very strongly scented flowers that are very dark in colour. Their dark colour would not attract bees or hummingbirds. In this case, a pollinator with poor vision and a very highly developed sense of smell is ideal. Bats possess these characteristics, and, not surprisingly, are the main pollinators of such plants, doing most of the pollen transfer at night.

So far, we have looked at how the colour and scent of plants attract specific pollinator species. Another factor to consider is the *structure* and shape of both the flower and the pollinator. Let's look at two examples of pollinators: butterflies and flies. Butterflies have long mouth parts that can reach nectar stored at the bottom of long tubular-shaped flowers. These "butterfly" flowers often have a convenient place for butterflies to land, so they can slurp the nectar with ease. Some flies, on the other hand, have the capacity to hover above a flower as if they were helicopters (hummingbirds have this capacity, too), and do not always need a landing pad. So flowers pollinated by these flies tend not to have landing pads.

THE COLOUR OF THIS FLOWER ATTRACTS ITS PREFERRED POLLINATOR, THE HUMMINGBIRD IN ARCADIA, CALIFORNIA, USA. THE FLOWER'S LONG TUBULAR SHAPE, PERFECT FOR A HUMMINGBIRD BILL, SUGGESTS THAT THE BIRD AND PLANT SPECIES EVOLVED TO RELY ON EACH OTHER FOR FOOD AND FOR POLLINATION.
© Danny Perez Photography

These examples show how the habits and physical characteristics of *both* the flower and the pollinator must be well-suited to each other for pollination to occur. Based on these characteristics, which types of animal pollinators visit which types of flowers in your neighbourhood?

THE WORLD'S LARGEST POLLINATOR

The ruffed lemur, a mammal found on the island of Madagascar, is the main pollinator of the traveller's tree (also called traveller's palm). These banana tree lookalikes are very tall, and can reach a height of 12 m. The lemur climbs the tree and, thanks to its nimble hands, opens the flower bracts and puts its long snout into the flower. In doing so its fur becomes covered in pollen. The pollen is transferred to the next traveller's tree flower that the lemur visits.

TRAVELLER'S TREE.
© Nolege

BLACK AND WHITE RUFFED LEMUR.
© Vision holder/Wikimedia Commons

CHAPTER 4 | Species: the cornerstone of biodiversity

CONSERVATION EFFORTS

Species are also important units in terms of conservation efforts. We often identify, prioritise and monitor biodiversity in terms of species, as we tend to understand them better than genes or ecosystems. As a result of the strong public interest in species, they also play a key role in engaging people in biodiversity conservation.

SEA OTTER
© Mike Baird/Wikimedia Commons

Let's take a look at a few terms used to describe different types of species that you might come across when learning more about conservation:

FLAGSHIP SPECIES: these are usually very charismatic, well-known species such as the giant panda or the tiger. Flagship species are used to help raise awareness of the need for conservation, by acting as mascots for all sorts of other species in need of our help.

UMBRELLA SPECIES: as a result of targeting conservation efforts towards one particular species, a whole host of other species might end up being protected. The target species is then often referred to as an umbrella species, as it provides cover for many others! For instance, by protecting an area of rainforest in order to conserve the beautiful jaguar, all other species that live within that habitat will also be protected.

KEYSTONE SPECIES: a keystone species is one which makes a disproportionately large contribution to the ecosystem it inhabits, given its biomass. Sea otters are keystone species; in terms of numbers, they do not form a massive part of the coastal area in which they live, yet, they make a huge contribution to their habitat. By eating sea urchins which, if left unchecked, can cause massive amounts of damage to their kelp forest habitat, sea otters help to maintain a balanced ecosystem for the other species living in the kelp forests.

THE STATUS OF SPECIES

Just as genes make up species, species make up ecosystems, which we will learn more about in the next chapter. Whether directly or indirectly, the survival of a species within an ecosystem often depends on the presence of several other species, and as such, the conservation of biodiversity is of utmost importance.

In Chapter 2 we had a look at some of the main causes of biodiversity loss, including habitat loss and fragmentation, overexploitation, climate change, invasive alien species and pollution. Each of these can place enormous pressure on species, leading to many being driven to extinction (see adjacent figure).

Extinction is a natural process which has been occurring since life on Earth began. There is a natural balance to life with the cycle of births and deaths of individuals. Over time some species thrive and evolution creates fascinating new species whilst others, unable to adapt to changing circumstances, become extinct.

Due to human activities the current rates of extinction are estimated to be 100 to 1 000 times higher than the normal background rate.

The problem we face today is not that extinction is occurring, but rather the *rate* at which it is happening. As outlined in Chapter 2, current rates of extinction are estimated to be 100 to 1 000 times higher than the normal background rate, due to human activities that are having a devastating effect on plant and animal life.

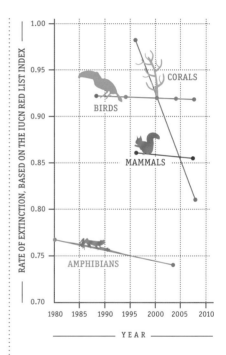

THE PROPORTION OF WARM-WATER CORAL, BIRD, MAMMAL AND AMPHIBIAN SPECIES EXPECTED TO SURVIVE WITHOUT ADDITIONAL CONSERVATION ACTIONS HAS DROPPED SINCE 1980.

CORAL SPECIES ARE MOVING MOST RAPIDLY TOWARDS GREATER EXTINCTION RISK. AMPHIBIANS ARE THE MOST THREATENED GROUP. THE IUCN RED LIST INDEX RANGES FROM 0 TO 1. A VALUE OF 0 MEANS ALL SPECIES IN A GROUP HAVE GONE EXTINCT. A VALUE OF 1 MEANS THAT ALL SPECIES IN A GROUP ARE NOT EXPECTED TO BECOME EXTINCT IN THE NEAR FUTURE.

Adapted from: Global Biodiversity Outlook 3, 2010

THE YOUTH GUIDE TO BIODIVERSITY

CHAPTER 4 | Species: the cornerstone of biodiversity

RED FOR DANGER...

Species already driven to extinction include the famous dodo, as well as the lesser-known hula painted frog, woolly-stalked begonia and pig-footed bandicoot. Unfortunately, there are thousands of species set to follow in their footsteps, all of which are in danger of being wiped out completely as a result of habitat destruction, pollution, overexploitation, climate change, invasive alien species or any combination of these.

A RECREATION OF THE EXTINCT DODO AND A SKELETON AT THE NATIONAL WATERFRONT MUSEUM IN WALES.
© Amgueddfa Cymru/National Museum Wales

THE IUCN RED LIST OF THREATENED SPECIES™

With so many species in need of conservation action, and with limited resources to help them, how do we know which are most at risk and most in need of our help? This is where the IUCN Red List of Threatened Species™ (also called the IUCN Red List) comes in.

The IUCN Red List is the world's most comprehensive information source on the global conservation status of species; it currently holds information about more than 48 000 different species, covering species taxonomy, geographic ranges, population numbers and threats. These data are collected by thousands of experts worldwide, and are an extremely useful tool in influencing conservation decisions, in informing species-based conservation actions, and in monitoring species' progress.

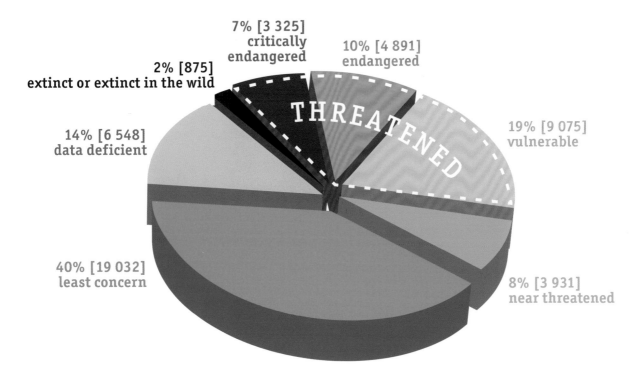

THIS FIGURE SHOWS THE PROPORTION OF SPECIES IN DIFFERENT THREAT CATEGORIES, WHICH REFLECT THE LIKELIHOOD THAT A SPECIES MAY BECOME EXTINCT IF CURRENT CONDITIONS PERSIST. THE RISK STATUS IS BASED ON THE RESEARCH OF THOUSANDS OF SPECIES DONE BY SCIENTISTS FROM AROUND THE WORLD. AS OF 2009, 47 677 SPECIES HAD BEEN ASSESSED. OF THESE, 36 PERCENT ARE CONSIDERED THREATENED WITH EXTINCTION.

Source: IUCN data in Global Biodiversity Outlook 3, 2010

Just as most hospitals have a triage system in place, whereby the ill and injured are assigned a category depending on how bad their condition is and therefore how quickly they need to be seen by a doctor, the IUCN Red List assigns species to special categories depending on how threatened they are.

There are eight categories for assessed species on the IUCN Red List, which can be seen on the scale accompanying the green turtle on the next page. A species is assigned a Red List category once its data have been assessed against very strict and carefully formulated criteria. These are based on factors such as geographic range, population size and rates of decline. Species that have been classified as Vulnerable, Endangered or Critically Endangered are referred to collectively as "threatened species".

CHAPTER 4 | Species: the cornerstone of biodiversity

NOT EVALUATED	DATA DEFICIENT	LEAST CONCERN	NEAR THREATENED	VULNERABLE	ENDANGERED	CRITICALLY ENDANGERED	EXTINCT IN THE WILD	EXTINCT
NE	DD	LC	NT	VU	EN	CR	EW	EX

By having this classification system and all the accompanying data, the IUCN Red List can help to answer several important questions, including:

- At what rate is biodiversity being lost?
- Where is biodiversity the highest?
- Where is it being lost most rapidly?
- What are the main reasons for these losses?
- How successful are conservation actions?

With these answers, conservationists and decision-makers are able to make more informed choices when developing and implementing conservation actions, therefore increasing their chances of success. With this success comes the preservation of biodiversity, which is so vital for our planet and all who live on it.

GREEN TURTLE.
© Kathryn Pintus

ALPINE IBEX: A CONSERVATION SUCCESS STORY!

The alpine ibex is endemic to Europe, and this once-abundant species used to roam freely across the Alps of France, Switzerland, Austria, Germany and northern Italy. However, as a result of intensive hunting practices, the alpine ibex was nearly driven to extinction in the early nineteenth century, with just a few hundred individuals remaining, all of which were found in the Gran Paradiso massif in Italy. Thanks to targeted conservation efforts, which included reintroductions to parts of its native range as well as introductions to Slovenia and Bulgaria, the alpine ibex is now listed as a species of Least Concern on the IUCN Red List, with a population of about 30 000 individuals recorded in the 1990s.

This species is not completely out of harm's way yet, though, as without continued efforts to protect its habitat, prevent poaching and reduce the impacts of human disturbance, it is likely to fall into a decline once again. The alpine ibex is thankfully not the only species that has been brought back from the brink of extinction, but it is a great example of what can be done to save species when we have the necessary knowledge and means.

ALPINE IBEX.
© Cash4Alex from Wikimedia Commons

CHAPTER 4 | Species: the cornerstone of biodiversity

CONCLUSION

In this chapter we looked at the importance of species diversity, particularly in relation to human livelihoods, and at the current status of the world's species. Despite the fact that the outlook seems less than bright, with many species currently at risk of extinction, there is still hope. There have been several success stories to date, with species being brought back from the brink of extinction through the careful application of conservation strategies. By implementing conservation tools such as the IUCN Red List, and using them to their full capacity to influence decisions and to inform action, biodiversity loss can be halted, if not reversed.

Most of the decisions that can be taken to instigate conservation efforts will be made by country leaders and officials, but they are not the only ones that can make a difference. We are all responsible for the well-being of our planet, and even the smallest of actions can have a positive effect.

There are plenty of things that each of us can do to help fight the extinction crisis:

:: Make informed decisions when considering what species of fish we eat, to help maintain wild fish stocks.

:: Be respectful towards wildlife, and only take part in ecotourism opportunities that are responsibly and ethically run, in order to prevent wildlife disturbance.

:: Recycle paper to reduce deforestation.

:: Spend a minute or two less in the shower each day to save water.

:: Use public transport to decre ase pollution levels, which may be contributing to global climate change.

The above are just examples of simple actions which you can implement in your daily lives. Think what specific actions you can undertake to contribute to the survival of species diversity.

LEARN MORE

:: Hunter Jr., M.L. (ed.) 2002. Fundamentals of Conservation Biology (2nd ed.) Blackwell Science, Inc., USA

:: Vié, J.-C., Hilton-Taylor, C. And Stuart, S.N. (eds.) 2009. *Wildlife in a Changing World – An Analysis of the 2008 IUCN Red List of Threatened Species.* Gland, Switzerland: IUCN. 180 pp.

:: The International Union for Conservation of Nature (IUCN): www.iucn.org

:: The IUCN Red List of Threatened Species: www.iucnredlist.org

:: ARKive: www.arkive.org

ECOSYSTEMS & ECOSYSTEM SERVICES

ECOSYSTEMS GIVE US FOOD, CLEAN WATER, CLEAN AIR, BALANCED HABITATS AND MUCH, MUCH MORE!

5

Nadine Azzu, FAO

POLLINATORS IN A GARDEN IN FRANCE.
© Richard Guerre (age 14)

An <u>ecosystem</u> can be considered the house where <u>biodiversity</u> lives – in terms of the physical location and the interactions that occur within this space. An ecosystem is made up of physical and chemical (abiotic) and living (biotic) factors – for example, rocks, air and water are physical/chemical factors, while plants, animals and micro-organisms are living factors.

CHAPTER 5 | Ecosystems and ecosystem services

An ecosystem is a **system** which contains biodiversity at all levels – including species diversity and genetic diversity – and encompasses the interactions and dependencies of biodiversity.

There are many reasons why different ecosystems are so fascinating – and one of those reasons is that a single ecosystem can contain many small ones. Let's take the case of a simple garden. In a garden, there can be grasses, flowers, bushes, maybe a tree or two, and if we want to be fancy, even a little pond. Of course, there is also the soil, and animals such as ants, worms and bees. But within that garden, there are what we can think of as microecosystems. For example, in the soil there are millions of tiny micro-organisms of all types. These micro-organisms are part of an intricate food chain, both under and above the ground. They also provide ecosystem services that keep the soil healthy, regulate water and capture carbon.

Ecosystems can be classified in various ways. Some ecosystems are *natural*, whereas others are *modified and managed by humans*. Ecosystems can be either *terrestrial* or *aquatic*. The different combinations of genes, species and microecosystems within an ecosystem are part of what makes each one unique.

One wonderful quality of ecosystems is their delicate balance. The abiotic and biotic factors interact with each other in such a way that all components of the ecosystem give and take just enough from each other, and in just the right way, to keep the ecosystem healthy. This "giving and taking" also allows for the ecosystem to provide different types of services (called ecosystem services) to the environment – including to humans.

TYPES OF ECOSYSTEMS

Terrestrial ecosystems are found on land, and include tropical forests and deserts. The biodiversity found in tropical rainforests is famous for its variety – birds of all types, shapes and colours, different and abundant tree species, and even spiders, snakes and monkeys.

Natural aquatic ecosystems can be inland or marine. Examples of natural freshwater ecosystems are ponds, rivers and lakes. The biodiversity found in a pond is very different from the biodiversity found, let's say, in a river. In a river, you could see salmon struggling and swimming upstream through the rushing waters to reach their breeding grounds. In a smaller, calmer pond, however, you might see ducks and fish swimming, water lilies floating along the surface of the water, insects flying overhead or frogs hiding in the shallows. Learn more about examples of freshwater biodiversity in Chapter 7.

Similarly, different marine ecosystems, such as seas, oceans and coral reefs, each contain their own unique biodiversity. Sharks, for example, can live in the open oceans, whereas corals, sponges and molluscs are more often found around sheltered coral reefs.

An <u>agro-ecosystem</u> is an example of an ecosystem that is dependent on human activities for its existence and maintenance. The biodiversity in agro-ecosystems provides food, fibre, medicine and other benefits for people. Examples of agro-ecosystems include rice paddies, pastures, agroforestry systems, wheat fields, orchards and even backyards with a homegarden or chickens (see the box: "The Rice Paddy Agro-ecosystem"). Read more about agricultural biodiversity in Chapter 9.

CHAPTER 5 | Ecosystems and ecosystem services

THE RICE PADDY AGRO-ECOSYSTEM

A rice paddy is an aquatic ecosystem that houses different types of fish, frogs, plants, insects and soil. For more than 5 000 years, humans have actively managed rice paddies to produce high yields of rice; these rice paddies are called flooded rice agro-ecosystems. In some countries, fish are kept in the rice paddy, so that farmers can harvest both rice *and* fish, which they eat and sell at the market. Similar to other ecosystems, a rice paddy *gives and takes*: when insects that come to eat the rice crop fall into the water, they become food for fish.

RICE TERRACES IN INDONESIA.
© FAO/Roberto Faidutti

WHAT ARE ECOSYSTEM SERVICES, AND WHY DO WE NEED THEM?

Ecosystem services (sometimes called ecosystem goods and services) are the benefits that the environment, of which humans are a part, obtains from ecosystems. Boxes: "The dirt on healthy soils" and "How our health and safety depend on biodiversity" take a closer look at some of the vital ecosystem services provided by biodiversity. There are four types of ecosystem services. They are:

1. PROVISIONING SERVICES: these services are products obtained from ecosystems, such as food, fresh water and genetic resources.

2. REGULATING SERVICES: regulating services are involved in climate regulation, disease control, erosion control, pollination and regulation of natural processes, such as floods and forest fires.

3. CULTURAL SERVICES: ecosystem services not only provide concrete things like food, or essential services such as water filtration, but they also provide us with spiritual, recreational and cultural benefits. For example, ecosystems provide a rich source of inspiration for art, folklore, national symbols, architecture and even advertising.

4. SUPPORTING SERVICES: these services maintain the conditions for life on Earth. They are necessary for the production of all other ecosystem services. Their impacts on people are either indirect or occur over a very long time. In contrast, changes in the other three categories have relatively direct and short-term impacts on people. Examples of supporting services are nutrient cycling, soil formation and retention and habitat provision.

1
FRESH FISH SOLD IN A MARKET IN CAMBODIA.
© World Bank/Masaru Goto

2
POLLINATION IS AN ECOSYSTEM SERVICE THAT DEPENDS TO A LARGE EXTENT ON THE COOPERATION, OR SYMBIOSIS, BETWEEN SPECIES - THE POLLINATED (THE PLANT) AND THE POLLINATOR (THE BEE). BEES PROVIDE VITAL POLLINATION SERVICES TO THOUSANDS OF PLANT SPECIES.
From Wikimedia Commons

3
BIODIVERSITY INSPIRED THESE ADVERTISEMENTS IN BONN, GERMANY DURING THE BONN INTERNATIONAL CONFERENCE ON BIODIVERSITY IN 2008.
© Christine Gibb

4
FORESTS PROVIDE VITAL HABITATS FOR MANY SPECIES.

CHAPTER 5 | Ecosystems and ecosystem services

THE DIRT ON **HEALTHY SOILS**

Enhancing the provision of ecosystem services can result from a collective effort. For example, creating and maintaining healthy soil in an agro-ecosystem requires farmers and worms to work together. Small <u>organisms</u> such as worms burrow through the soil and make it porous, so water can seep through, reaching the roots of plants. Worms also digest old leaves and plant material, recycling them into nutrients that nourish existing plants. In doing so, worms provide a very important ecosystem service – but they can't provide this service unless there is organic matter (old leaves and plant material) available. Humans also play a role in ensuring that soil is kept healthy and fertile. Farmers must decide carefully about which types of farming practices they use, so that the environment continues to provide ecosystem services. In an agro-ecosystem such as a crop field, farmers' practices such as <u>mulching</u>, leaving organic matter on the ground instead of collecting and disposing of it, provide the worms with the organic matter they will transform into nutrients to feed the farmers' crops.

CLOSE-UP OF WORM CULTURE. WORMS ARE USED TO IMPROVE SOIL QUALITY.
© FAO/A. Odoul

HOW OUR HEALTH AND SAFETY DEPEND ON BIODIVERSITY
Conor Kretsch, COHAB Initiative Secretariat

Biodiversity sustains our health in many ways. In addition to providing us with sources of fresh water and food, it provides important medicines and resources for medical research. Biodiversity also plays a role in the control of pests and infectious diseases, and by supporting healthy ecosystems it can help to protect us against the worst effects of natural disasters.

For around 80 percent of the world's people, healthcare is based on traditional medicines using wild flora and fauna. Many modern medicines are also based on chemical compounds from wildlife. The important anti-cancer drug Taxol comes from the Pacific yew tree and some types of fungus. The anti-malaria drug quinine comes from the cinchona tree, while the drug exanitide, which helps to treat diabetes, was developed from the venom of the gila monster lizard.

Modern medicine also has much to learn from studying animals in the wild. For example, wild bears eat large amounts of fatty and sugar-rich foods before hibernating for several months. In humans, eating fatty foods and sugars and not exercising for prolonged periods can lead to diabetes, obesity, heart problems and bone weakness; however, bears can sleep for 100 days or more without suffering from any of these problems!

So scientists studying bears hope to learn new ways of understanding and treating these diseases in people. Other species we are learning from include primates, crabs, sharks and whales. We still know very little about most of the biodiversity on Earth, but we know that when a species disappears, anything we might have been able to learn from that species disappears too.

A GILA MONSTER.
© Blueag9/Wikimedia Commons

CHAPTER 5 | Ecosystems and ecosystem services

A MOSQUITO RESTS ON FOLIAGE IN NAKHONRATCHASIMA PROVINCE, THAILAND.
© Muhammad Mahdi Karim/Wikimedia Commons

TSUNAMI HAZARD ZONE NOTICE ON A BEACH IN KOH PODA/ KRABI, THAILAND.
© Juergen Sack

Just as animals and plants have their own place and part to play in a healthy ecosystem, so too do organisms which can cause disease (e.g. certain viruses, bacteria, fungi and parasites). When human activity damages an ecosystem in which these organisms live, we risk creating new disease outbreaks. For example, the parasite that causes malaria is spread to people by the bite of some types of mosquito, which breed in pools of water. Changes to ecosystems – through deforestation, dam building or urbanisation – can provide new areas for mosquitoes to breed, and this can lead to an increase in malaria risk for people nearby.

Many other diseases have been linked to human impacts on biodiversity and ecosystems, including HIV/AIDS, SARS, hantavirus and some types of avian influenza.

Biodiversity also helps improve human safety and security. It can protect communities against the impacts of disasters, by supporting ecosystems that provide shelter against floods and storms, prevent erosion or avalanches on hillsides, or help provide food security for people faced with drought or famine. So, conserving biodiversity is a way of supporting communities and protecting our health and that of our children.

WHY SHOULD WE PAY ATTENTION TO ECOSYSTEM SERVICES?

As we have seen, balanced and healthy ecosystems provide important ecosystem services. They are important, not only for giving us clean air, water, soil and food, and for protecting us from floods and diseases, but also for providing us with beautiful landscapes in which to live. Ecosystem services are also vital to the short- and long-term survival and health of ecosystems. What other ecosystem services are there? What are some practical examples of why they are vital to humans and nature?

Human activity is putting ecosystems in danger, which in turn, means that these ecosystems cannot provide us (or any other part of biodiversity) with ecosystem services. But what exactly does this mean for us? And what can we do to help? Let's explore these questions using the ecosystem service of pollination as an example.

HONEYCOMB.
© Kriss Szkurlatowski

CHAPTER 5 | Ecosystems and ecosystem services

AN EXAMPLE OF AN ECOSYSTEM SERVICE:
POLLINATION

Pollination is an ecosystem service that depends to a large extent on cooperation, or symbiosis, between species – the pollinated (the plant) and the pollinator. Some pollinators only pollinate specific types of flowers. At least one-third of the world's agricultural crops (especially many fruits and vegetables) depend upon pollination provided by insects and other animals. Pollinators are essential for orchard, horticultural and forage production, as well as for the production of seeds for many root and fibre crops. Some examples of pollinators are moths, butterflies, flies, beetles and vertebrates (such as bats, squirrels and birds). Most animal pollination is done by bees. That means that bees are responsible for making sure many of the fruits and vegetables that we eat actually exist!

Bees visit flowers to drink nectar and to collect pollen grains. When a bee lands on the flower, pollen grains from the anther of the flower get stuck on the bee's body. Then, the bee flies off to another flower. Some of the pollen on the bee's body gets transferred to the stigma of this new flower – and in this way, the new flower is pollinated. Once a flower is pollinated, it produces a seed, and this seed can grow into a new plant.

A BEE COLLECTING POLLEN GRAINS.
© Laurence Packer/Cory Sheffield

LONG-HORNED BEE.
© John Hallmen/www.flickr.com

Unfortunately, bee populations are declining around the world. Many human practices kill bees, often by accident. For example, the uncontrolled spraying of pesticides kills both "bad" and "good" insects. The destruction of valuable bee habitats leaves fewer places for bees to live. Different bee species need different types of habitats for foraging and for shelter. The clearing of forest land harms species that live in hives or inside fallen logs. The ploughing of fields destroys the homes of ground-nesting bees.

Although scientists do not yet know all the reasons why bee populations are declining, they do know that the decline will have a huge impact on ecosystems and on our food. If pollinator populations fall, it will be difficult to grow crops that provide us with important vitamins and nutrients, such us our fruits and vegetables. Without diverse nutritious fruits and vegetables, we would end up having unbalanced diets and health problems.

Often, bees have a bad reputation – they are seen by people as being dangerous and irritating, and are generally not welcome. Instead, we should learn to appreciate the importance of pollination, and maybe help give bees a better reputation! So next time you are in a garden and see bees buzzing around, try to notice if there are any fruit trees around. Tell your family and friends that bees should not be seen as harmful insects, because if they ate a piece of fruit this morning for breakfast, it was thanks to the bees who pollinated the trees!

AN EXHIBIT AT THE EXPO OF DIVERSITY IN BONN, GERMANY SHOWS THE IMPORTANCE OF BEES IN PRODUCING FOOD. WITH BEES, WE HAVE ALL THE FOODS ON THE LEFT TABLE. WITHOUT BEES (RIGHT TABLE), WE HAVE MUCH LESS FOOD!
© Christine Gibb

CHAPTER 5 | Ecosystems and ecosystem services

CONCLUSION

In a sense, nature is made up of "big life" and "small life". The big life cannot survive if small life is not taken care of, or if it is not managed properly. The bits and pieces of small life – and just as importantly, the interactions between them - are what sustain the big life. The small life consists of the species that can be found on the ground, in the skies, in the water and underground – for example, mammals, birds, fish and insects. The big life is the wider ecosystem. As we have seen in this chapter through examples such as rice paddy ecosystems, pollination and soil fertility, it is the small life (species) that not only sustains the ecosystem but also the ecosystem services that ensure a healthy and functioning planet.

LADYBIRD INVASION.
© Tobias Abrahamsen (age 16)

We can draw an important lesson from these examples: in nature, we must look at both the individual "small life" and the "big life". In practical terms, this means that if we were doctors, and we saw a population of any given species suffering, we would not just prescribe a specific medicine to target the apparent illness – instead, we should find out why the species population is sick. Maybe the answer doesn't lie within the species itself, but is caused by a series of events found in the wider ecosystem. Can you give an example of how a natural or human-made impact on "small life" affects "big life"?

Together, we can take many actions to raise our own awareness, and that of others, of the importance of healthy ecosystems. We can start by taking small, but significant, actions. Begin, for example, by building a terrarium (small container with soil for plants) to learn firsthand how an ecosystem works. Monitor and record the activities that you observe in your terrarium. Take the terrarium to school and share your observations with your classmates. Whatever you decide to do, make sure you apply your learning about healthy ecosystems to your daily actions.

TERRESTRIAL BIODIVERSITY - LAND AHOY!

EXPLORING LAND-BASED ECOSYSTEMS, FROM DEEP IN THE SOIL TO HIGH IN THE MOUNTAINS

6

Saadia Iqbal, Youthink!, The World Bank

As you know by now, <u>biodiversity</u> can be classified in three different ways (refer to Chapter 1). In Chapter 5 we learned that one way to classify it is by <u>ecosystems</u>, which includes the important category of terrestrial biodiversity. It might sound a bit creepy, but guess what: you're part of it! And so am I. That's right – unless you happen to be a fish, or maybe some super-intellectual seaweed (in which case you really should be on TV), you count as a member of this group. So who are they (uh, we) anyway?

A PRAYING MANTIS IN GREECE.
© Jonas Harms (age 20)

CHAPTER 6 | Terrestrial biodiversity - land ahoy!

TYPES OF TERRESTRIAL BIODIVERSITY

Terrestrial biodiversity refers to animals, plants and micro-organisms that live on land, and also land habitats, such as forests, deserts and wetlands.

Not so creepy after all! At least... not always.

Terrestrial biodiversity is mind-bogglingly vast. Each animal and plant species, and the ecosystem in which it lives, has a unique contribution to our world, and plays a part in keeping the delicate balance of things intact. Let's take a closer look at some of the different kinds of terrestrial biodiversity, why they are important and the threats they are facing.

A DOMINICAN LIZARD.
© Chad Nelson

A BLACK AND WHITE JUMPING SPIDER.
© Godfrey R. Bourne/nsf.gov

YOUTH AND UNITED NATIONS GLOBAL ALLIANCE

FORESTS

Forests are one of the Earth's greatest treasures – rich habitats teeming with animal and plant species, herbs, fungi, micro-organisms and soils. They provide people with food, wood, medicine, fresh water and clean air, and millions of the world's poorest people rely on forests for their livelihoods. To say that forests help to nurture all life on the planet is no exaggeration.

FOREST IN AZERBAYJAN.
© FAO/Marzio Marzot

WELCOME TO THE JUNGLE!
Forests are home to about 80 percent of the world's land-based animals and plants. Around 300 million people around the world live in forests!

Forests also influence nature's capacity to cope with natural hazards. Their destruction could cause altered rainfall patterns, soil erosion, flooding of rivers and the potential <u>extinction</u> of millions of species of plants, animals and insects.

As if all this weren't enough, forests are also huge storehouses of carbon, which means they absorb carbon from the atmosphere and convert it into plant tissue. This is very important for reducing the impact of <u>climate change</u>, a change in the overall state of the Earth's climate caused by a build-up of greenhouse gases in the Earth's atmosphere. Widespread deforestation, therefore, may increase global warming.

IT'S A FACT!
The Canadian boreal forest stores an estimated 186 billion tonnes of carbon, which is equal to 27 times the world's carbon emissions in 2003 from the burning of fossil fuels.

Source: International Boreal Conservation Campaign

THE YOUTH GUIDE TO BIODIVERSITY

CHAPTER 6 | Terrestrial biodiversity – land ahoy!

WHAT'S THREATENING FORESTS?

Although trees are a renewable resource that can replenish themselves, they are being cut down faster than they can grow back. There are many factors behind this problem, for example:

- **Clearing forested land to grow crops:** for many poor people, it is a tricky situation. They cut down trees to meet their short-term needs, but in the long run, they lose their forests, and therefore their livelihoods, as a result of deforestation. Economic incentives often convince forest owners to sell their land, cut down forests, and grow export items such as coffee and soybeans. However, the once-forested land is often poor in nutrients unless it is managed very carefully, so farmers are only able to use it for a few years before they must move to a different area of the forest and clear it for farming. Sometimes the abandoned area is used to raise livestock, but it takes 2.4 hectares of pastureland in the tropics to feed just one cow. That's the size of six football pitches! You can see that raising livestock in tropical rainforests is not very sustainable!

- **Cutting down trees for wood:** people need wood for many reasons including for fuel, building homes and making furniture. When individuals or timber companies cut down trees in an irresponsible way, the process can harm the surrounding areas and wildlife. Illegal logging is also a big problem.

- **Other threats** to forests include mining, settlements and infrastructure development. Climate change may increase the impacts of pests and diseases. Climate change is also predicted to result in more extreme climatic events in many places, such as through floods and droughts, which will harm forest plant and animal populations and could cause more wildfires. Also, changes in rainfall and temperature will force species to migrate – which may not be possible if there is no suitable habitat for them or if they are slow moving (or, in the case of a tree, can't move at all). Climate change also alters the phenology of many species (the timing of biological events such as flowering and fruiting).

© FAO/Roberto Faidutti

RAINFORESTS

Rainforests can be either temperate or tropical. Both kinds have a few things in common: high rainfall all year round, and very lush, dense and tall vegetation. Both are also rich in plant and animal species, although the diversity is greater in tropical rainforests. Tropical rainforests are warm and moist; while temperate rainforests are cool and moist.

Because of human activities, rainforests are disappearing at an alarming rate: a few thousand years ago, over 15.5 million square kilometres of tropical rainforest existed worldwide. Today, only 6.7 million square kilometres are left. Not only is this a terrible loss to the Earth's natural beauty and diversity, it will also hurt people's lives and well-being.

Tropical rainforests contain more biodiversity than any other ecosystem on Earth. They cover less than two percent of the Earth's total surface area, and are home to 50 percent of the Earth's described plants and animals! Like all forests, rainforests play a huge role in reducing atmospheric carbon dioxide levels.

Also, with their millions of plants, they regulate temperatures through a process called transpiration, in which plants return water to the atmosphere. Transpiration increases humidity and rainfall, and has a cooling effect for miles.

Another way in which the rainforest provides tremendous value to the world is through its medicinal plants. It is estimated that one out of four ingredients in our medicine comes from rainforest plants; and so far, less than one percent of the tropical rainforest species has been analysed for their medicinal value!

URUCAN SEEDS FROM THE BRAZILIAN AMAZON.
THE SEEDS ARE USED BY LOCALS TO PRODUCE A RED POWDER THAT IS MIXED WITH VEGETABLE OIL AND APPLIED TO THEIR SKIN, LIKE SUNSCREEN, TO PROTECT IT FROM THE SUN. IT IS ALSO USED AS AN INSECT REPELLENT.
© Igor Castro da Silva Braga/World Bank

CHAPTER 6 | Terrestrial biodiversity – land ahoy!

MOUNTAINS

Mountains don't just *look* big and tall – their contribution as ecosystems is sky high too! They supply fresh water to almost half of the world's population, and on every continent (except Antarctica), they provide mineral resources, energy, forest and agricultural products. Vegetation on mountains provides a range of environmental benefits.

> **FOOD FOR THOUGHT...**
> *Of the 20 plant species that supply 80 percent of the world's food, six (maize, potatoes, barley, sorghum, tomatoes and apples) originated in mountains!*

For example, it influences the water cycle by capturing moisture from the air. Snowfall in the high mountains is stored until the snow melts in the spring and summer, providing essential water for settlements, agriculture and industries in the surrounding lowlands. Mountain vegetation helps to control this water flow, preventing soil erosion and flooding. Vegetation on mountains also helps to reduce climate change through carbon storage. Many of the herbs, game and other foods that sustain people are found on mountains.

MOUNTAINS IN CANADA.
© Curt Carnemark/World Bank

WHAT'S THREATENING MOUNTAINS?

LIVING THE HIGH LIFE
Mountains cover 25 percent of the Earth's land surface. They are home to 12 percent of its people.

Mountain areas are facing a loss of diversity due to a number of factors, including the uphill expansion of agriculture and human settlements and logging for timber and fuelwood.

The illegal trade in animal parts and medicinal herbs is also contributing to biodiversity loss on mountains. Climate change is another factor that is threatening several species with extinction. Many plant species are moving uphill, partly due to climate change, which reduces available land area for those <u>organisms</u> already living there and increases competition for space and other important resources. Also, those species which already live at the top of mountains can't move further up to get to colder conditions.

Watch a slideshow about species migration in the Amazon:
digitalmedia.worldbank.org/SSP/youthink/amazon

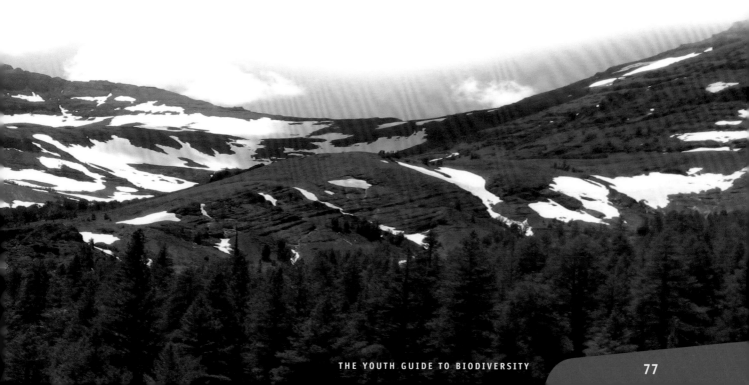

THE YOUTH GUIDE TO BIODIVERSITY

CHAPTER 6 | Terrestrial biodiversity - land ahoy!

SOIL
BIODIVERSITY

It might not sound like an exciting place to live, but you'd be surprised how many living things chose soil as their home! Soil contains a myriad of organisms, such as earthworms, ants, termites, bacteria and fungi. In fact, a typical handful of garden soil contains *billions to hundreds of billions* of tiny soil micro-organisms!

Together, soil organisms contribute with a wide range of services to their ecosystems, such as improving the entry and storage of water, preventing erosion, improving plant nutrition, and breaking down organic matter. In addition, soil biodiversity influences the environment indirectly in many ways. For example, it helps regulate pest and disease occurrence in agricultural and natural ecosystems, and can also control or reduce environmental pollution. Soil is the second biggest carbon storehouse, after forests, with some soils, such as peat, actually storing more carbon than forests on a hectare-by-hectare basis.

WOMAN HOEING IN TANZANIA.
© Scott Wallace/World Bank

DID YOU KNOW? Antibiotics such as penicillin and streptomycin are derived from soil organisms.

78 YOUTH AND UNITED NATIONS GLOBAL ALLIANCE

WHAT'S THREATENING SOIL BIODIVERSITY?

The diversity of soil organisms is under threat from pollution, unsustainable agriculture, overgrazing, vegetation clearing, wildfires and poor management of irrigation. Converting grasslands or forests to cropped land results in rapid loss of soil carbon, which indirectly enhances climate change. Urbanisation and soil sealing – covering of land for housing, roads or other construction work – are also threats, because concrete ends up killing the life present in the soil beneath.

There are many other types of terrestrial biodiversity, including biodiversity in dry and sub-humid lands (see box: "Drylands Biodiversity") and in wetlands (see Chapter 7). They each play an important part in our ecosystems' healthy and productive functioning, and keep our Earth diverse and beautiful.

A DOWN TO EARTH LIFESTYLE…
Plant roots are also soil organisms because of their beneficial symbiotic relationships *and interactions with other soil dwellers. They prevent soil erosion and help drain water from the soil, keeping it from becoming too wet. Conversely, they can help to hydrate the soil when it becomes too dry. Roots also help to make soil, by splitting rocks into tiny pieces that eventually become soil.*

CHAPTER 6

DRYLANDS BIODIVERSITY
Jaime Webbe, CBD

WILDEBEEST OR GNU IN THE DRYLANDS OF KENYA.
© Curt Carnemark/World Bank

Dry and subhumid lands, also known as <u>drylands</u>, cover about 47 percent of the Earth's land surface and include everything from deserts to savannas to Mediterranean landscapes. Although drylands are commonly imagined to be barren, dead landscapes, they contain a number of important, well-adapted species. The Serengeti grasslands in sub-Saharan Africa, for example, support an annual migration of approximately 1.3 million blue wildebeests, 200 000 plains zebra and 400 000 Thomson's gazelles. The rocky, shrub-dominated Mediterranean Basin in Europe and North Africa, contains 11 700 <u>endemic</u> plant species that are unique to these areas.

Biodiversity in dry and subhumid lands is critical for human survival. For example, some of the world's most important food crops originated in drylands including wheat, barley and olives. Dryland biodiversity also provides genetic sources for one third of the plant-derived drugs available in the United States alone. Finally, the traditional knowledge associated with livelihoods in drylands, including from pastoral peoples, is important for <u>sustainable development</u>, a long-term process of enlarging people's choices and freedom.

Unfortunately, the biodiversity of dry and subhumid lands is facing a number of threats from human activities. Between six and 12 million square kilometres of dry and subhumid lands are affected by <u>desertification</u> – that is, the degradation of land to such an extent that production is reduced. Already at least 2 311 species are threatened or endangered in drylands, while at least 15 species have disappeared completely from the wild. This trend shows no sign of reversing as drylands are among the most vulnerable regions to the negative impacts of climate change. In sub-Saharan Africa, for example, between 25 and 40 percent of mammals in national parks may become Endangered while as many as two percent of the species currently classified as Critically Endangered may become Extinct as a result of climate change.

Given the challenges faced by the biodiversity of dry and subhumid lands, it is important to take action now. We need to learn more about these important regions and the value of their biodiversity in providing critical <u>ecosystem services</u>. We need to involve native or <u>indigenous peoples</u> living in drylands in decision-making. We need to address the global challenges of climate change and desertification.

PROTECTED AREAS

Protected areas are places that receive protection because of their environmental or cultural value. They have many purposes, including the **conservation** and sustainable use of biodiversity. Most countries have protected areas. There are over 100 000 protected sites around the world, covering around 12 percent of the Earth's land surface.

■ UNDER 10%   ■ 10 - 30%   ■ 30 - 50%   ■ OVER 50%

AT LEAST 10 PERCENT OF OVER HALF THE WORLD'S 825 ECOREGIONS ARE LISTED AS PROTECTED AREAS. THE LIGHTER COLOURING ON THE MAP REPRESENTS ECOREGIONS WITH RELATIVELY LOW LEVELS OF PROTECTION.
Source: UNEP-WCMC in GBO-3, 2010

Well-managed protected areas support healthy ecosystems, which in turn keep people healthy. Globally, protected areas meet millions of people's most basic needs by providing essentials such as food, fresh water, fuel and medicines both for the people living in and around protected areas and even for people living hundreds or even thousands of kilometres away. They also benefit local communities by promoting rural development, research, conservation, education, recreation and tourism. Protected areas can also act as buffers against climate change and poverty, and, of course, they are reservoirs of biological richness for present and future generations.

> *"If you were to randomly remove parts from a computer or car, everyone knows that both those systems will become less reliable or very likely stop working altogether. The same thing happens to ecosystems when they lose their species."*
>
> **Shahid Naseem**
> Director of Science at the Earth Institute's Center for Environmental Research and Conservation.

CHAPTER 6 | Terrestrial biodiversity - land ahoy!

WHAT CAN YOU DO?
HERE ARE A FEW THINGS YOU CAN DO TO HELP PROTECT TERRESTRIAL BIODIVERSITY:

- Learn about biodiversity in your community. Which plants and animals are native to your area? Are they facing any threats?

- Help protect natural areas and "green spaces" in your communities, even ones as small as the neighbourhood park.

- Try to buy locally grown and organic fruit and vegetables, when possible, but also remember that products sustainably produced in the developing world are important for people's income and livelihoods.

- Buy products from certification schemes, which guarantee that certain environmental and social principles were followed in producing the product. Some examples are the Forest Stewardship Council, Marine Stewardship Council and Fair Trade.

- Help keep your environment clean and beautiful; keep an eye out for litter, and choose household products (cleaners, paints, etc.) that do not contain any pollutants.

BEACH CLEAN-UP.
© Danil Nenashev/World Bank

:: Ask your parents to avoid using pesticides in your garden. Start a compost pile to reduce waste in your home and to help fertilise the soil in your garden.

:: Reduce your carbon emissions by turning off unused lights, switching to fluorescent light bulbs in your home, and using public transportation or walking and cycling whenever possible. Eating more veggies will help too!

:: Spread the word! Bug your friends, teachers, siblings and parents. Ask them to take these steps in their lives too. Together, we can all keep our Earth healthy, beautiful and full of life!

LEARN MORE

:: **Amazon Conservation Association:** www.amazonconservation.org
An organisation dedicated to preserving biodiversity in the Peruvian and Bolivian Amazon.

:: **Conservation International - Biodiversity Hotspots:** www.biodiversityhotspots.org/Pages/default.aspx
Explore the places on Earth that are the richest in plant and animal life—and also the most threatened.

:: **Forest Graphics:** grida.no/_res/site/file/publications/vital_forest_graphics.pdf
Learn about the different types of forests in the world and where they are located.

:: **International Boreal Conservation Campaign:** www.interboreal.org/globalwarming
Read about the boreal forest.

:: **The Nature Conservancy:** www.nature.org
Find out about conservation projects around the world.

:: **World Database on Protected Areas:** www.wdpa.org
Discover the world's protected areas.

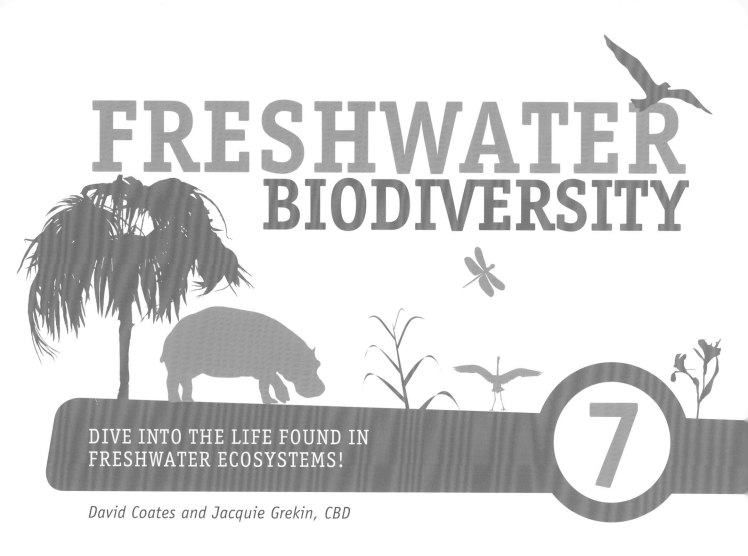

FRESHWATER BIODIVERSITY

DIVE INTO THE LIFE FOUND IN FRESHWATER ECOSYSTEMS!

David Coates and Jacquie Grekin, CBD

Freshwater includes rivers, lakes and wetlands and are habitats which are rich in biodiversity. Such systems provide us with many services such as our drinking water, food (such as fish), a means of transport as well as recreational opportunities. Unfortunately freshwater systems are some of the most endangered habitats in the world and have alarming rates of species extinction.

LAKE CHAD BASIN, AFRICA.
© FAO/Marzio Marzot

CHAPTER 7 | Freshwater biodiversity

WHAT ARE FRESHWATER ECOSYSTEMS?

Simply put, "fresh" water is water without salt, which distinguishes these environments from marine, or saltwater, <u>ecosystems</u>. There are many kinds of freshwater ecosystems, such as:

RIVERS: in which the water flows, usually towards the sea.

LAKES: larger areas of standing water (shallow or deep).

WETLANDS: areas of land covered either permanently or temporarily with water, usually shallow, covered by plants (including trees) which grow out of the water or mixed with areas of open water. Examples of wetlands include swamps, marshes, bogs, <u>peatlands</u>, estuaries, mangroves and rice paddies.

Freshwater ecosystems are part of the landscape and interact with land. For example, rainwater falling on land flows into streams and rivers, and fills up lakes and wetlands, carrying with it nutrients and plant material (such as seeds and leaves).

But freshwater ecosystems also supply water to land environments – for example, they provide water to recharge water stored below ground (groundwater), which supports plants living on land (such as forests). These movements of water are part of the "water cycle" (see box: "The Water Cycle"), which connects land, "groundwater", freshwater and coastal areas.

SHOTOVER RIVER CANYONS IN QUEENSTOWN, NEW ZEALAND.
© Alex E. Proimos/flickr.com

LAKE IN CHILE.
© Curt Carnemark/World Bank

MANGROVES IN THE GALAPAGOS ISLANDS.
© Reuben Sessa

THE WATER CYCLE

The water cycle is the continous movement of water around the planet. During this cycle water can be in various states: solid, liquid or gas. Water moves by processes of evaporation (water turning from a liquid to a gas), transpiration (the movement of water through vegetation and soil), condensation and precipitation. Water travels above and infiltrates below the ground and accumulates in rivers, lakes and oceans and evaporates or transpires into the atmosphere where it condenses to form clouds and then returns to the Earth's surface through precipitation (rain, snow, hail and sleet). The changes in the state of water during the cycle requires the exchange of heat, therefore cooling or heating the environment (for example, evaporation requires energy and therefore cools the environment). The water cycle also has the effect of purifying water courses, replenishing water supplies and moving nutrients and other elements to different parts of the world.

Biodiversity (i.e. trees and other plants) is a necessary part of the cycle. The soils in which they are rooted absorb water and store it safely, while their leaf canopies return water, in the form of vapour, to the atmosphere, where it becomes precipitation. Large-scale removal of vegetation can disturb the cycle, often resulting in changed rainfall patterns and soil erosion. Biodiversity, therefore, supports the availability of water for people and other living things to use.

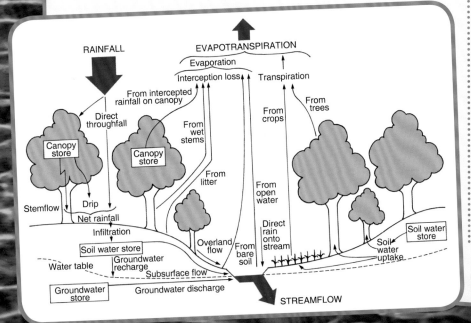

Graphic source: L. S. Hamilton 2008. Forests and Water. FAO Forestry Paper 155, Rome: FAO, 3.

CHAPTER 7 | Freshwater biodiversity

FRESHWATER LIFE

"Freshwater biodiversity", at the species level, includes life which is very obviously living in freshwater but also includes life which is adapted to live in or around freshwater habitats. Examples include:

- Fish
- Amphibians (e.g. frogs and salamanders)
- Wetland-dependent mammals (e.g. hippopotamuses (see box: "The Hippopotamus"), river dolphins (see box: "River Dolphins: Species in Danger"), porpoises, seals, otters, moose, beaver, manatees)
- Waterbirds (e.g. pelicans, flamingos, cranes, ducks, geese)
- Reptiles (e.g. crocodiles, turtles)
- Insects (e.g. dragonflies, mosquitoes)
- Aquatic plants and plants rooted in water but with stems and leaves that emerge from the water

There are also many plants which are adapted to life in or near freshwater habitats, other than those permanently living underwater. This includes peat, sedges (tall grass-like plants, including papyrus), mangroves and rice (see box: "Rice Paddies: Farmed Wetlands").

PELICANS AT THE LAKE NAKURU NATIONAL PARK IN KENYA.
© Thérèse Karim

THE **HIPPOPOTAMUS**

The hippopotamus is considered to be a freshwater mammal because, although it feeds on land, like a cow, it is adapted to life in water.

For example, it has a flat head on which the eyes and nostrils protrude, enabling it to remain submerged, but still able to see and breathe; a cow could not do this.

A TREE GROWING IN THE MEKONG RIVER IN LAOS – NOTE HOW THE TREE ROOTS BUTTRESS THE TREE AGAINST THE RIVER FLOW (RIGHT TO LEFT IN THIS PHOTO), AN ADAPTATION TO LIFE IN RIVERS.
© David Coates

HIPPOS SWIM IN LAKE NAIVASHA, KENYA.
© Véronique Lefebvre

FLAMINGOS AT THE LAKE NAKURU NATIONAL PARK IN KENYA.
© Thérèse Karim

CHAPTER 7 | Freshwater biodiversity

RIVER DOLPHINS: SPECIES IN DANGER

RIVER DOLPHIN.
© Dolf En Lianne

CHINA'S BAIJI, OR YANGTZE RIVER DOLPHIN.
© Cathy McGee

Although often regarded as marine (saltwater) species, some dolphins live exclusively in freshwater rivers and lakes; others have adapted to live in both marine and freshwater environments. River dolphins differ from oceanic dolphins in several ways, including having a much longer snout – up to 20 percent of their body length – and extremely poor eyesight. Most are comparable in size (about 2.5 m) to the more common and better-known bottlenose dolphin, a marine species seen in aquariums and featured in movies and on television.

There are six species of river dolphins:

:: Ganges River dolphin (Bangladesh, India, Nepal, Pakistan) – the "Susu"
:: Indus River dolphin (Pakistan)
:: Amazon River dolphin (South America) – the "Boto"
:: Yangtze River dolphin (China) – the "Baiji"
:: Irrawaddy and Mekong River dolphin (salt- and freshwater – Myanmar, Laos and Cambodia)
:: Tucuxi (salt- and freshwater – east coast of Central and South America).

The Yangtze River dolphin has been presumed extinct since 2006; the others (with the possible exception of the Amazon River dolphin and the Tucuxi, about which little data are available) are highly endangered. By comparison, the bottlenose dolphin is relatively abundant, and is not in danger of extinction.

The survival of river dolphins is threatened by habitat loss and degradation, as a result of dam construction and river diversion, which reduces water flow; pollution from industry and agriculture; overfishing; and accidental capture in fishing lines and nets (known as bycatch).

RICE PADDIES:
FARMED WETLANDS

Rice is a wetland-dependant (freshwater) plant and the staple food for over half the world's population. It provides about 20 percent of the total calorie supply in the world and is grown in at least 114 countries worldwide, particularly in Asia.

Rice paddies are naturally flooded or irrigated fields in which rice is grown. Rice grows with its roots submerged, but with its leaves and seeds (rice) above the water. Rice paddies usually dry out at harvest time, illustrating that these systems shift between aquatic and terrestrial (dry land) phases.

Rice is just one crop. But living in the water in the paddy fields are thousands of species of aquatic **organisms**. Rural populations benefit directly from some of this biodiversity by harvesting reptiles, amphibians, fish, crustaceans, insects and molluscs for household consumption. But other biodiversity associated with rice paddies supports the health and productivity of the rice itself through, for example, controlling rice-pests and helping to make nutrients available to the rice plants.

These wetlands also support the conservation of internationally important populations of resident and migratory waterbirds.

TERRACED RICE PADDIES NEAR A RED ZAO VILLAGE OUTSIDE OF SAPA, LAO CAI PROVINCE, NORTHERN VIETNAM.
© Tran Thi Hoa/World Bank

CHAPTER 7 | Freshwater biodiversity

THE IMPORTANCE OF FRESHWATER BIODIVERSITY

ORANGUTANS LIVE IN PEAT SWAMP FORESTS AND ARE THREATENED AS A RESULT OF HABITAT LOSS. THEY ARE <u>endemic</u> TO THE INDONESIAN ISLANDS OF BORNEO AND SUMATRA AND ARE FOUND NOWHERE ELSE IN THE WORLD (EXCEPT IN CAPTIVITY). THIS BORNEAN ORANGUTAN HELPS TO SPREAD TREE SEEDS, INCLUDING SOME SPECIES THAT CAN ONLY GERMINATE WHEN THEY HAVE PASSED THROUGH THE ORANGUTAN'S GUT!
© Borneo Orangutan Survival Foundation

Freshwater biodiversity provides a variety of benefits (<u>ecosystem services</u>) to people, including:

Food: in developing countries, inland fisheries can provide the primary source of animal protein for many rural communities (see box: "Aquaculture"). Aquaculture, the farming of fish and other aquatic animals (e.g. shrimp), can also provide food and income for many people, as can wetland agriculture, such as rice farming.

Fibre: throughout human history, many wetland plants have been a source of fibre for making such items as baskets, roofs, paper and rope. Papyrus, for example, was used for making paper as early as 4000 BC (think of the ancient Egyptian scrolls).

Recreational and cultural benefits: many rivers, lakes and wetlands are highly valued for recreational and cultural benefits, some of which have high economic value (such as tourism). In developed countries, sport fishing is also an important recreational activity and a significant source of income for many communities. Recreational fishers have been a major driving force in cleaning up freshwater environments to restore recreational benefits.

AQUACULTURE

Aquaculture is the farming of fish and other aquatic animals and plants (e.g. shrimp, frogs, mussels, oysters and seaweed). Freshwater aquaculture can be very beneficial and provide food and income for many people, particularly in rural communities in developing countries.

Aquaculture originated as freshwater carp farming in Asia and is now widespread. Asia still leads the way in this industry, accounting for 92 percent of global production (70 percent in China, 22 percent in the rest of the region).

Worldwide, about half of production is in fresh or brackish water (a mix of fresh and salt water), and the other half in marine environments. Most aquaculture production in freshwater is fish. The main freshwater species cultured include carp, tilapia, pacu, catfish and trout.

Production from freshwater fish species tends to be more sustainable than from marine species because most of it is based on vegetarian rather than carnivorous species. For example, it can take two kilograms of fish to produce one kilogram of salmon (a carnivorous fish), which doesn't sound like a good deal. Better to eat lower down the food chain!

Aquaculture can cause water pollution (from chemical use and waste products) and introduce invasive alien species (species that have spread outside of their natural habitat and threaten biodiversity in the new area). Efforts must be made to address these impacts, particularly as aquaculture develops, expands and intensifies.

A FISH FARM NEAR NEW DELHI, INDIA.
© FAO

CHAPTER 7 | Freshwater biodiversity

Carbon storage: climate change is largely due to the release of carbon dioxide and other greenhouse gases into the atmosphere. Wetlands, particularly peatlands, are "carbon sinks": they remove and store significant quantities of carbon from the atmosphere. Peatlands alone store more than twice as much carbon as all the world's forests. Destruction of these wetlands results in the release of carbon into the atmosphere, increasing the intensity of global climate change. Human exploitation has destroyed 25 percent of the peatlands on Earth.

Water purification and filtration: plants, animals and bacteria in forests, soils and wetlands also filter and purify water. Wetland plants accumulate excess nutrients (such as phosphorus and nitrogen) and toxic substances (such as heavy metals) in their tissues, removing them from the surrounding water and preventing them from reaching drinking water. They can be thought of as "nature's kidneys" (see box: "Biodiversity = Clean Water = Human Health").

Flood regulation: many wetlands provide a natural flood barrier. Peatlands, wet grasslands and floodplains at the source of streams and rivers act like sponges, absorbing excess rainwater runoff and spring snowmelt, releasing it slowly into rivers and allowing it to be absorbed more slowly into the soil, preventing sudden, damaging floods downstream. Coastal freshwater-dependent wetlands, such as mangroves, saltmarshes, tidal flats, deltas and estuaries, can limit the damaging effects of storm surges and tidal waves by acting as physical barriers that reduce the water's height and speed. As global climate change raises sea levels and increases extreme weather in many parts of the world, the need for these services has never been greater.

SITTING ON A ROOF, RESIDENTS IN NEW ORLEANS WAIT TO BE RESCUED AFTER HURRICANE KATRINA.
© Jocelyn Augustino/FEMA

BIODIVERSITY = CLEAN WATER = HUMAN HEALTH

All life depends on water. Human beings need two to three litres of clean drinking water a day. Without food we can survive weeks. But without water, we can die of dehydration in as little as two days. More than one billion people in the world lack access to safe drinking water, and some two million people die each year of diarrhoea caused by unclean water, 70 percent of these are children.

Healthy ecosystems contribute to providing clean water supplies. Many cities, for example, obtain their water supply from protected areas outside the cities.

WATER SAMPLES FROM INDIA'S MUSI RIVER TAKEN AT INTERVALS UP TO 40 KM DOWNSTREAM OF HYDERABAD. ON THE LEFT, CLOSE TO THE CITY, WATER IS HIGHLY POLLUTED FROM UNTREATED WASTES. WATER QUALITY IMPROVES DOWNSTREAM AS THE ECOSYSTEM BREAKS DOWN THIS WASTE.
© Jeroen Ensink

CHAPTER 7 | Freshwater biodiversity

THREATS TO FRESHWATER BIODIVERSITY

Biodiversity is being lost more rapidly in freshwater ecosystems than in any other ecosystem type.

- Some 20 percent of freshwater fish species are considered extinct or threatened, a far greater percentage than for marine fishes.
- 44 percent of the 1 200 waterbird populations with known trends are in decline (compared to 27.5 percent of seabirds being threatened).
- 42 percent of amphibian species populations are declining.
- Among groups of animals that live in many different areas, those species living in association with freshwater tend to show the greatest level of threat (including, for example, butterflies, mammals and reptiles).
- On average, over half of the natural wetland area has probably been lost in most developed countries. In Canada, for example, more than 80 percent of the wetlands near major urban centres have been converted for agricultural use or urban expansion; in many others, the loss is higher than 90 percent (e.g. New Zealand).

This loss of biodiversity is because of the human demands placed on freshwater and wetland habitats due to such factors as:

- **Conversion of habitat**, through the draining of wetlands for agriculture, urban development or damming of rivers.
- **Overuse** of water for irrigation, industrial and household use, interfering with water availability; (agricultural production alone accounts for over 70 percent of water extracted from rivers – the biggest use of water worldwide).
- **Pollution** of water through excess nutrients (phosphorus and nitrogen) and other pollutants such as pesticides and industrial and urban chemicals (see box: "Aren't Nutrients Good For You?").
- **Introduction of alien species**, causing local extinction of native freshwater species.

These threats are rapidly increasing as human populations grow and demands on water escalate.

Climate change is also becoming an important threat to wetlands and their biodiversity. Its main impacts will be on fresh water: melting glaciers and ice-caps (which are fresh water) causing rising sea levels, and changes in rainfall (less of it in some areas, leading to drought, more of it in others, leading to excessive flooding). One projection indicates that water availability will decrease in about a third of the world's rivers. Almost half the world's population will be living in areas of high water stress by 2030.

AREN'T NUTRIENTS GOOD FOR YOU?

What's wrong with nutrients? Aren't they good for you? All living things need nutrients, such as nitrogen and phosphorus, to grow and survive. That's why these nutrients are the main ingredients in agricultural fertilisers (helping crops to grow). Excess nutrients are also contained in sewage from both households and farms (excreted from all living things).

The problem arises when nutrients are dumped untreated or washed into waterways in excessive quantities: this leads to the excess growth of certain plants (algae), which consume the oxygen in the water as they grow and decay. This process, known as "<u>eutrophication</u>", makes the water unliveable for fish, and the algal blooms make the waterways unpleasant for recreational use; in some cases algal blooms even become poisonous.

CHAPTER 7 | Freshwater biodiversity

WHAT CAN BE DONE?

A number of organisations and international agreements aim to protect freshwater biodiversity, including:

- The Convention on Biological Diversity: this Convention has a programme of work specifically dedicated to protecting inland waters biodiversity.

- The Ramsar Convention on Wetlands: is an intergovernmental treaty that guides national action and international cooperation for the conservation and sustainable use of wetlands and their resources; almost 1 900 "Wetlands of International Importance" have been designated under the treaty.

- Wetlands International: a global organisation that works to sustain and restore wetlands and their resources for people and biodiversity.

- The International Union for Conservation of Nature (IUCN), The Nature Conservancy (TNC), The World Wide Fund for Nature/World Wildlife Fund (WWF) and Conservation International (CI) all have freshwater programmes. There are many other non-governmental organisations (NGOs) addressing freshwater issues at regional, national and local levels.

FIND OUT WHERE YOUR WATER COMES FROM...

The first step in protecting freshwater biodiversity is to become aware of where fresh water comes from and how much we depend on it: not just for what we drink, but for personal hygiene, growing our food, and producing energy and the goods we consume.

FIND OUT HOW MUCH WATER YOU DRINK. AND EAT.
AND WEAR. AND DRIVE. AND…

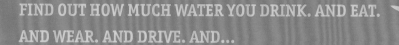

Globally, people use an average of 633 cubic metres per year.

Water footprints however vary greatly from one part of the world to another; for example, the average person consumes 173 cubic metres per year in sub-Saharan Africa, 581 cubic metres in Europe, and 1 663 cubic metres in North America.

Of the water consumed, only about 0.75 to 1.5 cubic metres per year, much less than one percent, is actually used for drinking. We consume much more in other ways, particularly by eating it.

Here are some water requirements to produce typical products:

Hamburger: 2 400 litres
Glass of milk: 200 litres
Cup of coffee: 140 litres
Cup of tea: 35 litres
Glass of apple juice: 190 litres
Cotton T-shirt: 4 100 litres
Pair of leather shoes: 8 000 litres
Tonne of steel: 230 000 litres

Meat production in particular, especially beef, consumes a great deal of water. The average volume of water (worldwide) required to produce one tonne of beef is 15 497 cubic metres; compare this to a tonne of chicken (3 918) or a tonne of soybeans or barley (1 789 and 1 388, respectively).

A sustainable diet, anyone?

CLEAN HOME, CLEAN EARTH...

Another way to reduce your impact on waterways is to reduce or eliminate your use of chemicals. Many laundry detergents today are phosphate-free, but this is not the case for most dishwasher detergents. What about the other cleaning, personal hygiene and gardening products you use? Are they really necessary? Find out what they contain and how you can replace them: for instance, there are plenty of biodegradable alternatives for many of the products we typically use. Most garden chemicals can be avoided by changing the plants grown, gardening practices and accepting a more natural landscape (which can also look nicer).

LOOK UPSTREAM, DOWNSTREAM AND BENEATH YOUR FEET...

Want to get more involved? Look "upstream" – and see how sustaining the water catchment can improve water security. Look "downstream" – and see how you can reduce your impact. And don't forget to look beneath your feet – promote the conservation of groundwater by avoiding polluting or overusing it and maintaining the nature above ground that helps to replenish it.

Join a group – or start one – and help clean up rivers and lakes, including the banks and wetlands. Support wetland protection and restoration. Promote approaches to water supply and management that use the abilities of ecosystems to supply clean water more securely and for reducing flood risk.

THE GOOD NEWS...

The loss of freshwater biodiversity and degradation of ecosystems are not necessarily irreversible. For example, many countries, in both rich and poor regions, are starting to take steps to restore wetlands that were drained in the relatively recent past. This is being done because the benefits of restoring the services provided by these wetlands can outweigh the costs of not having those services (e.g. poor water quality and increased flood risk). The process begins with public recognition of the values of these ecosystems to people and the economic benefits of managing them more wisely.

LEARN MORE

- Conservation International (CI): www.conservation.org
- Hamilton 2008. Forests and Water. FAO Forestry Paper 155, Rome: FAO, 3.
- The International Union for Conservation of Nature (IUCN): www.iucn.org
- The Nature Conservancy (TNC): www.nature.org
- Peatlands: www.wetlands.org/Whatwedo/PeatlandsandCO2emissions/tabid/837/Default.aspx
- The Ramsar Convention on Wetlands: www.ramsar.org
- Water Footprints: www.waterfootprint.org/Reports/Report16Vol1.pdf earthtrends.wri.org/pdf_library/data_tables/wat2_2005.pdf
- Wetlands International: www.wetlands.org
- The World Wildlife Fund / The World Wide Fund for Nature (WWF): www.wwf.org

THE RICHES OF THE SEAS

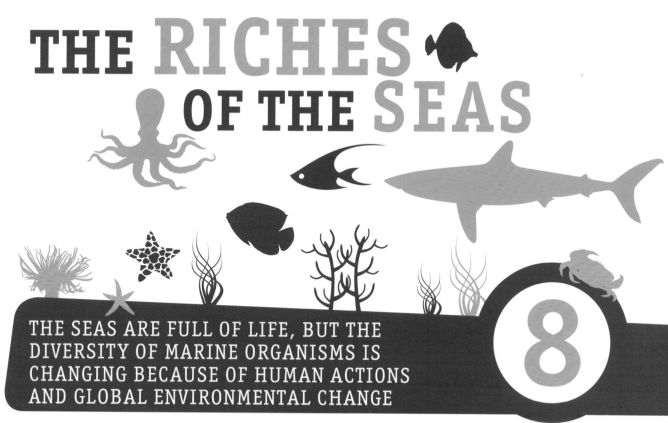

THE SEAS ARE FULL OF LIFE, BUT THE DIVERSITY OF MARINE ORGANISMS IS CHANGING BECAUSE OF HUMAN ACTIONS AND GLOBAL ENVIRONMENTAL CHANGE

Caroline Hattam, Plymouth Marine Laboratory

Did you know that life began in the ocean some 3.5 billion years ago? And did you know that scientists estimate that there may be up to 10 million <u>species</u> living within the seas?

The marine environment is home to a stunning variety of beautiful creatures, ranging from single-celled <u>organisms</u> to the biggest animal ever to have lived on the Earth – the blue whale.

This chapter describes the multitude of life found in the seas, the uses we make of it and how marine life is changing because of this use and because of global environmental change.

GALÁPAPOS CRAB.
© Reuben Sessa

CHAPTER 8 | The riches of the seas

THE UNDERSIDE OF A HORNED STARFISH, STARFISH ISLAND, PALAWAN, PHILIPPINES.
© Vince Ellison B. Leyeza (age 12)

EASTERN SPINNER DOLPHINS IN THE EASTERN TROPICAL PACIFIC.
© William High/NOAA Fisheries

MICROSCOPIC ORGANISMS (RADIOLARIA) UNDER POLARISED LIGHT. THESE FOSSILS WERE FOUND IN BARBADOS.
© Biosphoto/Gautier Christian

A GREEN TURTLE IN HAWAII, USA.
© Mila Zinkkova/Wikimedia Commons

MARINE LIFE

We know much less about marine <u>biodiversity</u> than we do about terrestrial biodiversity, but we do know a few interesting facts:

- There are 35 animal <u>phyla</u> (groups of animals such as arthropods and molluscs) found in the seas, 14 of which are only found in the sea.
- The marine environment is home to both the largest mammal on the Earth (the blue whale) and the biggest invertebrate (the colossal squid).
- The largest marine mammals are often dependent on the smallest marine life for food. For example, blue whales feed on krill. Krill are small animals that weigh about one gram each, and a blue whale needs to eat about 3.6 million of them every day!
- The fastest animal in the sea is the sailfish which can reach speeds of 100 km/h (imagine someone swimming at the same speed as your car the next time you are on the highway).
- The oldest known living marine creature is a deep water black coral found off the coast of Hawaii and is estimated to be over 4 000 years old!

BLUE PARROTFISH
© Rosaria Macri

CHAPTER 8 | The riches of the seas

GIANTS OF THE DEEP

The marine environment is home to a number of giant animals, for example:

:: The **blue whale** is known to grow to over 30 m and can weigh as much as 181 tonnes (the same as almost 20 cars)!
:: **Giant clams** can grow to over a metre in size and may live for over 100 years.
:: The **colossal squid** is even bigger than the giant squid. Colossal squids weigh about the same as a small cow (about 500 kg) and can measure over 10 m.
:: **Giant isopods**, distant relatives of garden woodlice (also known as pill bugs), can grow to over 30 cm in length.

Many of these giant creatures grow very slowly and take many years to mature and produce offspring. This makes them very susceptible to human activities and environmental change as they are slow to adapt.

BLUE WHALE AND A SCUBA DIVER.
© brianlean.wordpress/www.flickr.com

GIANT CLAM.
© Ewa Barska/Wikimedia Commons

GIANT DEEP SEA ISOPOD IN TE PAPA MUSEUM, NEW ZEALAND.
© Y23/Wikimedia Commons

COLOSSAL SQUID, GULF OF MEXICO.
© NOAA

MARINE HABITATS

Coastal areas are very productive and support large numbers of marine organisms. The number of marine organisms is often greatest in relatively shallow coastal areas because they are rich in nutrients and light. Many of these nutrients, which are food for marine life, come from the land. Some coastal areas are extremely diverse, for example coral reefs (see box: "Coral Reefs").

CORAL REEFS

Coral reefs are one of the most diverse ecosystems on the planet, containing very high numbers of marine species. Scientists have so far described 4 000 reef fish and 800 coral species. Coral reefs are also important for people and they provide income, food and a livelihood for more than 500 million people, mostly in developing countries. Corals are very sensitive to changes in sea temperature and there are fears that global warming will cause the death of many coral reefs.

GREAT BARRIER REEF, AUSTRALIA.
© Rosaria Macri

THE YOUTH GUIDE TO BIODIVERSITY

CHAPTER 8 | The riches of the seas

THE OPEN OCEAN CONTAINS SMALL POPULATIONS OF VERY MANY DIFFERENT SPECIES

The open ocean has few nutrients available, so despite its enormous size, it is not home to dense populations of organisms, but the diversity of these organisms is very high. Here you find trillions of small single cell organisms known as phytoplankton (e.g. diatoms, dinoflagellates and coccolithophores) and larger zooplankton (e.g. copepods and foraminifera). You also find many kinds of fish and whales.

Only a very small amount of light passes below 100 to 200 m, and no light reaches beyond 500 to 1000 m. This environment far below the surface is very stable, cold and dark. Many of the organisms that live in this part of the ocean have evolved special adaptations that help them to survive in this environment. For example, some organisms swim up to the upper zones of the oceans to feed at night. Others have developed special body parts called **photophores** which are bioluminescent (they produce light). Some fish have developed enormous mouths with very sharp teeth and their jaws can be unhinged to catch big prey.

THE SEA BED PROVIDES AN IMPORTANT HABITAT FOR MARINE LIFE

We know much more about the creatures that live on the sea bed than we do about some of those that live in the water, mostly because the bottom-dwellers don't move very quickly and are easier to catch! In shallow, light areas, it is possible to find marine plants (e.g. seagrasses) and algae (or seaweeds) which look like plants but actually are not very closely related. In and on the seabed you can also find starfish, sea urchins, polychaete worms, sea cucumbers, anemones, sponges, corals and shelled animals such as clams, mussels and scallops... the list is almost endless.

DIATOMS.
© C. Widdicombe/PML

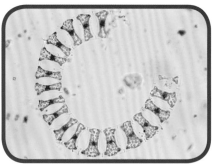

DIATOMS.
© E. Fileman/PML

DINOFLAGELLATES.
© C. Widdicombe/PML

COCCOLITHOPHORES.
© PML

COPEOPOD.
© PML

THE DEEP SEA

The deep sea is full of weird and wonderful life. It is not a flat, barren place but contains a number of [biodiversity hotspots](), areas of high species diversity and habitat richness, such as:

Seamounts: these underwater mountains provide a range of living conditions suitable for rich and diverse marine communities.

Cold-water coral reefs: found from 200 to 1000 m below the surface, they provide food and shelter for hundreds of different species, including commercially important fish.

Deep sea sponge fields: these are found in clear, nutrient-rich waters and provide a living space for many invertebrates and fish.

Hydrothermal vents: are found in volcanically-active areas where warm, mineral-rich water is released into the sea. The food chain is based on bacteria that convert sulphur compounds into energy. The bacteria support large numbers of organisms.

Gas hydrate vents and cold seeps: these are areas on the sea floor where hydrocarbons and mineral-rich cold water escape into the sea. The bacteria found here use methane to produce energy and, like hydrothermal vents, they support large communities of animals.

CHAPTER 8 | The riches of the seas

HOW WE USE MARINE BIODIVERSITY

Humans are dependent upon marine biodiversity in more ways than you think!

When you think of how we use marine life, you probably think of fish and shellfish for food. Although the oceans are important sources of food, they also provide many other important benefits, such as:

From left to right:
MANGROVES, GALAPAGOS.
© Reuben Sessa

COASTAL RECREATION, LYME REGIS, UK.
© S. Boyne

STUDENTS ON PRAPAS BEACH, THAILAND STUDYING TAXONOMY.
© PML

EXAMINING MICROSCOPIC MARINE ORGANISMS IN THE LAB.
© scalefreenetwork/www.flickr.com

FISHING TRAWLER SURROUNDED BY GREAT CRESTED TERNS.
© Marj Kibby/www.flickr.com

- **The balance of our climate:** some marine organisms (e.g. phytoplankton) take up carbon dioxide from the atmosphere. Others produce gases, such as dimethyl sulphide, which can help form clouds that reflect the sun's rays and cool the planet.

- **The breakdown and removal of waste and pollution:** bacteria in sea water can break down organic waste (e.g. sewage), some can even breakdown petrochemicals and have been used to help clean up oil spills. Larger marine animals eat organic and inorganic materials (e.g. metal compounds) and can bury them within the sea bed.

- **The reduction of damage from storms:** the presence of salt marshes, coral reefs, mangroves and even kelp forests and seagrass meadows can reduce the amount of energy in waves, making them less destructive when they reach the shore.

- **Recreation:** millions of people use the marine environment for recreation and many are drawn to it because they can see marine life (e.g. dolphins, whales, sea birds, seals and manatees). Coral reefs are also a popular tourist attraction and are estimated to generate US$ 9.6 billion worldwide for the tourism industry.

- **Learning experiences:** some schools and youth groups take young people on field trips to the beach to learn about marine life. Has your school taken you?

> **SEAWEED SURPRISES**
> *Did you know that you've probably already eaten some seaweed today? It's in your toothpaste! And maybe you've put some on your face and hair as it's found in many shampoos and cosmetics such as creams and lotions. It's also used as a fertiliser, animal feed, in medicines, gums and gels. Maybe you've used it in a science lesson at school as a medium called agar on which to grow bacteria. Perhaps in the future you'll also be using it to fuel your car as scientists are already using marine algae to produce biofuels.*

- **New medicines, biofuels and other products:** many pharmaceutical and biotechnology companies study marine life in the search for new compounds that may be useful for people. So far more than 12 000 potentially useful compounds have been found in marine organisms.

- **Our heritage and culture:** the sea and sea life appear in many folk tales, novels, poetry, songs and works of art. Can you think of any?

- **Our health and well-being:** many people find being near the sea to be relaxing and inspirational, and seeing marine life adds to the enjoyment. Doing exercise on the beach or in the sea is being promoted as a way to improve our health and well-being.

THE YOUTH GUIDE TO BIODIVERSITY

CHAPTER 8 | The riches of the seas

THREATS TO MARINE BIODIVERSITY

Marine biodiversity faces a number of threats that are causing changes in the mixture of species, where they are found, and in some cases, <u>extinction</u>. The IUCN lists 27 percent of corals, 25 percent of marine mammals and over 27 percent of seabirds as threatened. Particular threats include over-fishing (see box: "Fishing Down the Food Chain"), <u>pollution</u> (see box: "Pollution and Dead Zones"), <u>climate change</u> and ocean acidification and <u>invasive alien species</u>.

Climate change is leading to changing ocean temperatures, which in turn, cause species to migrate. In the northern hemisphere, some cool water species are moving north, while in the southern hemisphere cool water species are moving south. Warm water species are expanding their distribution into areas where cool water species once lived.

Climate change is also affecting the pH (or level of acidity) of the sea. As more carbon dioxide (CO_2) enters the air, more CO_2 is absorbed by sea water through a natural chemical reaction. This is causing the sea to become more acidic. The full effects of this are not known, but scientists think that it will affect the building of shells or calcium-rich structures (such as corals) and reproduction in many species.

The combined effects of over-fishing, pollution and climate change are making it much easier for non-native species to establish themselves in new areas. Many of these alien species are not problematic, but some are; they are said to be invasive. Invasive alien species can be very difficult to eliminate of and may out-compete native species causing the whole ecosystem to change. Ships are the main culprits in the spread of invasive alien species, transporting them unintentionally on their hulls or in their <u>ballast water</u>.

FISHING DOWN THE FOOD CHAIN

Most of the fish we prefer to eat are large, slow-growing species (e.g. cod, tuna and snapper). As their numbers are falling, fishers are changing the species that they catch. They are increasingly catching smaller fish (e.g. mackerel and sardines) that are closer to the bottom of the food chain. These smaller fish may be the prey of the larger fish, so removing these smaller fish threatens the recovery of the larger fish. An example is cod fishing in Norway. As cod numbers decreased, fishers started targeting pout. Pout feeds on krill and copeopods. Krill also feed on copeopods, as do young cod. As the pout were caught, krill increased causing copeopods to decrease. The young cod then found it difficult to find food, making the recovery of the cod population even more difficult.

POLLUTION AND DEAD ZONES

Pollution enters the marine environment through a number of routes. It may come from ships as they move around the oceans, from the land (e.g. from industrial outlets, sewage outfalls and runoff from roads) and from rivers. It includes rubbish, sewage and many different chemicals, such as fertilisers, oil and medicines.

Pollution carried in rivers is becoming particularly problematic and in some cases, where river water containing high levels of fertilisers reaches the coast, it is leading to "dead zones". Dead zones are becoming much more common in coastal waters around the world. About 200 have been identified so far. Some come and go with the seasons, but others are permanent. The most well-known is found in the north of the Gulf of Mexico where the Mississippi River meets the sea. At its largest it covered 22 000 km².

Dead zones appear when fresh water carrying lots of nutrients meets the sea. The fresh water floats on top of the salty water and prevents oxygen moving down. In the spring and summer, phytoplankton grow and multiply rapidly because of the high level of nutrients. Some of the phytoplankton are difficult to digest and may produce poisonous substances. This means less food is available for other marine life. As the phytoplankton die, they fall to the seafloor and are broken down by bacteria. This process needs oxygen. As no oxygen can get through to the seabed, the oxygen already there gets used up quickly and the seabed is said to become hypoxic (almost no dissolved oxygen is present). Marine animals and plants, like those on land, need oxygen to survive. Those that can move leave the area, but those that cannot are left to die.

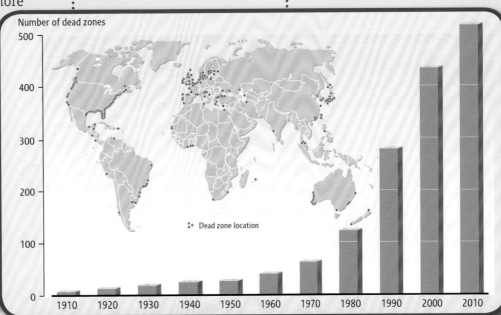

THE NUMBER OF MARINE DEAD ZONES HAS INCREASED OVER THE DECADES.
Source: GBO-3, 2010

CHAPTER 8 | The riches of the seas

WHAT IS BEING DONE?
Save the seas and we'll save the planet

Internationally a number of pieces of legislation aim to protect marine biodiversity, for example:

- The Ballast Water Management Convention which, when it comes into action, will aim to reduce the introduction of invasive alien species by ships.

- The Convention on Biological Diversity aims to protect all biodiversity, including that found in the oceans.

- International fisheries legislation and codes of conduct (e.g. FAO's Code of Conduct for Responsible Fishing and the European Union's Common Fisheries Policy) aim to encourage sustainable fishery management.

Individual countries are also acting:

- Increasing numbers of marine protected areas are being created by many countries around the world, but currently only 0.7 percent of the marine environment is protected globally.

- Aquaculture or fish farming is being encouraged as an alternative to wild capture fisheries and new approaches to aquaculture may improve its environmental footprint (see box: "Multi-Culture Aquaculture").

Local people are also making a difference:

- They are getting involved in beach cleaning, coastal surveys and pollution campaigns.

- They are reducing their use of plastic bags that often end up being swept out to sea.

- They are encouraging sustainable fishing by only buying fish that carries the Marine Stewardship Council logo.

MULTI-CULTURE AQUACULTURE

Instead of just farming one species of fish, some fish farmers are diversifying to include shellfish (e.g. mussels) that can filter out organic waste from the water (e.g. fish faeces) and seaweeds to use up the excess nutrients that leak out of fish farms. The fish farmers can then sell not only the fish, but also the shellfish and the seaweed!

Perhaps you could do the same.

:: Do a project at school about some of the things you have learned here, or even get your school to organise a beach clean-up.

:: Make biodiversity friendly choices when buying products or eating food derived from oceans, for example by selecting certified productions, such as those with Certified Sustainable Seafood labels (www.msc.org).

:: Most importantly, you could spread the message about how important marine biodiversity is and how we need to protect it.

LEARN MORE

If you would like to find out more about marine biodiversity, try these links:

:: The European Environment Agency's 10 Messages for 2010: Marine Ecosystems: www.eea.europa.eu/publications/10-messages-for-2010/message-4-marine-ecosystems.pdf
:: Global marine biodiversity trends: www.eoearth.org/article/Global_marine_biodiversity_trends#
:: The IUCN's marine programme: www.iucn.org/about/work/programmes/marine
:: Marine biodiversity: www.eoearth.org/article/Marine_biodiversity#
:: UNEP-WCMC's report Deep-sea biodiversity and ecosystems: A scoping report on their socio-economy, management and governance: www.unep-wcmc.org/resources/publications/UNEP_WCMC_bio_series/28.aspx
:: World database on marine protected areas: www.wdpa-marine.org/#/countries/about
:: World register of marine species (WoRMS): www.marinespecies.org/about.php

IN FARMERS' FIELDS: BIODIVERSITY & AGRICULTURE

AGRICULTURAL BIODIVERSITY IS CRUCIAL FOR PRODUCING FOOD AND OTHER AGRICULTURAL PRODUCTS. USING IT ENHANCES FOOD SECURITY AND NUTRITION, AND HELPS FARMERS ADAPT TO CLIMATE CHANGE

9

Ruth Raymond and Amanda Dobson, Bioversity International

Like water and air, <u>agricultural biodiversity</u> is a basic resource that we literally could not live without. But the importance of agricultural biodiversity is not well understood and so it is not properly valued. The result is that agricultural biodiversity is under threat both in the wild and in farmers' fields. This is a threat we cannot afford.

BUGS ON A SUNFLOWER IN THE USA.
© Alex Sorensen (age 14)

CHAPTER 9 | In farmers' fields: biodiversity & agriculture

WHAT IS AGRICULTURAL BIODIVERSITY?

Agricultural biodiversity includes the different ecosystems, species and genetic variability that contribute to food production. Some components of agricultural biodiversity, such as livestock breeds and crop varieties, are actively managed by farmers and scientists. Others, such as soil microbes and many pollinators, provide valuable services without being actively managed.

The variation within plant and animal species enables them to evolve and adapt to different environmental conditions.

Farmers and professional breeders depend heavily on agricultural biodiversity, which allows them to develop plant varieties and livestock breeds that can resist pests and diseases, can adjust to changing climates, and have higher nutritional value.

WHAT DOES BIODIVERSITY HAVE TO DO WITH AGRICULTURE?

Agriculture depends on the diversity of relatively few plant and animal species. Approximately 250 000 plant species have been identified, 7 000 of which can be used as food. But only 150 crops are cultivated on any significant scale worldwide and only three (maize, wheat and rice) supply nearly 60 percent of the protein and calories in the human diet.

Given its heavy dependence on just a few food species, humanity relies on the diversity within these species to survive. This diversity can be considerable. For example, there are tens of thousands of different varieties of rice, developed by farmers over millennia. The International Rice Research Institute in the Philippines holds about 110 000 samples of different rice varieties in cold storage.

THE THREE MAIN FOOD CROPS WORLDWIDE:

MAIZE
© Curt Carnemark/World Bank

WHEAT
© Britta Skagerfalt/Global Crop Diversity Trust

RICE
© IRRI

THE YOUTH GUIDE TO BIODIVERSITY

CHAPTER 9 | In farmers' fields: biodiversity & agriculture

Crop varieties may differ in plant height, yield, seed size or colour, nutritional qualities or flavour. They may respond differently to cold, heat or drought. Some varieties have the ability to withstand pests and diseases that would prove fatal to others.

In the wild, biodiversity is the result of **natural selection**: the evolution of animals and plants to meet the challenges of their environment. In the field, this is the result of thousands of years of human activity. In addition, tremendous agricultural biodiversity has been created through the careful selection of useful traits by farmers, plant breeders and researchers. Today, modern biotechnology is changing the way agriculture is done (see box: "Biosafety and Agriculture").

The use of **biodiversity** is the key to productive agriculture. Farmers continually require new plant varieties that can produce high yields under different environmental circumstances without large amounts of fertilisers and other agrochemicals. Crop diversity provides farmers and professional plant breeders with options to develop, through selection and breeding, new and more productive crops that are nutritious and resistant to pests and diseases.

Livestock farmers also need a broad gene pool to draw upon if they are to improve the characteristics of their animals under changing conditions. Traditional breeds, suited to local conditions, survive times of drought and distress better than exotic breeds and, therefore, frequently offer poor farmers better protection against hunger.

Bananas are the fourth most important crop after rice, wheat and maize. If all the bananas grown in the world each year were placed end to end, they would circle the Earth 1 400 times!

A BANANA TREE IN THE DEMOCRATIC REPUBLIC OF THE CONGO.
© Strong Roots

The global agricultural labour force includes approximately 1.3 billion people, about a fourth (22 percent) of the world's population and nearly half (46 percent) of the total labour force.

Source: MILLENNIUM ECOSYSTEM ASSESSMENT, 2005

THE MULTIPLE BENEFITS OF AGRICULTURAL BIODIVERSITY

> Agricultural biodiversity is a resource that is available to everyone. In fact, some of the world's poorest countries are the wealthiest in terms of agricultural biodiversity.

Agricultural diversity underpins dietary diversity, which contributes to lower mortality, greater longevity and a decrease in diseases normally associated with affluence, such as obesity, heart disease and diabetes.

Agricultural biodiversity can improve <u>agricultural productivity</u> without costly inputs. Another benefit of agricultural biodiversity is somewhat intangible and difficult to quantify, but not less important. It relates to the sense of national pride and identity that arises when people come to understand the value of their traditional native foods.

Yet another advantage of agricultural biodiversity is that it buffers yields. Total harvests may be lower in a diversified production system, but they are more stable from year to year. This suits small farmers in rural areas, who seek to minimise risk – ensuring that there will always be some food for their families – rather than to maximise productivity.

AGRICULTURAL DIVERSITY IS A SOURCE OF NATIONAL PRIDE AND IDENTITY. BREAD SELLER IN TAJIKISTAN.
© Gennadiy Ratushenko / World Bank [www.flickr.com/photos/worldbank/4249171838]. page 121

LOCAL MARKET IN KYRGYZ.
© Nick van Pragg/World Bank

AGRICULTURE DIVERSITY HELPS TO MINIMISE RISK. DIFFERENT POTATO VARIETIES CAN GROW UNDER DIFFERENT CONDITIONS, AND CAN BE USED FOR A VARIETY OF DISHES. IN SOME COMMUNITIES IN THE HIGH ANDES, FARMERS WILL GROW FOUR OR FIVE SPECIES OF POTATOES ON A SMALL PLOT OF LAND. MANY OF THESE FARMERS STILL MEASURE THEIR LAND IN 'TOPO', THE AREA A FAMILY NEEDS TO GROW THEIR POTATO SUPPLY. THE SIZE OF A TOPO VARIES: IT IS LARGER AT HIGHER ALTITUDES BECAUSE THE LAND NEEDS MORE TIME TO LIE FALLOW AND RECOVER BETWEEN PLANTINGS COMPARED TO LOWER ALTITUDES.
© INIAP

CHAPTER 9 | In farmers' fields: biodiversity & agriculture

BIOSAFETY AND AGRICULTURE
Ulrika Nilsson, CBD

For over 10 000 years, farmers have selected and saved their best seeds and animals for breeding so that future generations of plant varieties and animal breeds would have better qualities in terms of size, taste, growth rate or yield. In recent years, new techniques, called modern biotechnology, have allowed scientists to modify plants, animals and micro-organisms at rates faster than they can with conventional methods.

Scientists can take a single gene from a plant or animal cell or from a bacterium and insert it in another plant or animal cell resulting in a [living modified organism](#) (LMO). LMOs are commonly known as genetically modified organisms (GMOs), even though LMOs and GMOs have different definitions. Although modern biotechnology can potentially improve human well-being, for example by improving agricultural productivity, there is concern about possible risks that LMOs may pose to biodiversity and human health.

In response, world leaders adopted the Cartagena Protocol on [Biosafety](#), a supplementary agreement to the Convention on Biological Diversity. The Protocol works to protect biodiversity by encouraging the safe transfer, handling and use of LMOs. It does so by establishing rules and procedures for regulating the import and export of LMOs from one country to another. As of October 2011, 161 Parties (160 countries and the European Union) have adopted the Protocol.

The Protocol describes two main procedures. One is for LMOs intended for direct introduction into the environment, such as live fish and seeds. It is called the Advance Informed Agreement (AIA) procedure. The other is for LMOs used for food, feed or processing (LMOs-FFP), such as tomatoes. Under the first procedure, countries must assess if the LMOs could pose any risks. Based on the results of the risk assessment, a country can decide whether or not to import the LMO.

Under the second procedure, countries that approve LMOs-FFP for domestic use and placement on the market must inform other countries and provide relevant information through a central information exchange mechanism known as the Biosafety Clearing-House (BCH). If a country decides to import an LMO that is to be released into the environment, it must communicate its decision and a summary of the risk assessment to the BCH. Countries must also make sure that LMOs shipped from one country into another are safely handled, transported and packaged. Shipments of LMOs must be accompanied by documents that clearly specify their identity and any requirements for their safe handling, storage, transport and use.

If you're interested in biosafety issues, you can encourage your government to become a Party to the Protocol if it is not already, inform others about biosafety issues, discuss possible public educational activities on biosafety with your teachers, design awareness material on biosafety to use in your community, or create a youth biosafety network to exchange information.

THE FIRST COMMERCIALLY GROWN LMO, A TOMATO MODIFIED TO RESIST ROTTING, WAS INTRODUCED IN 1994.
© FAO/Olivier Thuillier

CHAPTER 9 | In farmers' fields: biodiversity & agriculture

THREATS TO AGRICULTURAL BIODIVERSITY

With the advent of modern agriculture, untold numbers of locally adapted crop varieties were replaced by genetically uniform, high-yielding modern varieties. In China, for example, between 1949 and 1970, the number of wheat varieties grown by farmers dropped from about 10 000 to 1 000. Farmers in India once grew 30 000 rice varieties.

Today, 75 percent of India's rice crop comes from just 10 varieties. Only 20 percent of the maize varieties known in Mexico in 1930 can be found there now. Overgrazing, climate change and changes in land use are also taking a toll on the diversity still available in the <u>wild relatives of crops</u> and other wild species.

<u>Extinction</u> is a natural process. Species have emerged, flourished, and died out over the ages. What is alarming is that, due mainly to human activity, today's rate of extinction is thousands of times greater than the rate at which new species appear. This threatens the life-sustaining systems of our planet.

In the past 100 years, as much as 75 percent of the genetic diversity of agricultural crops may have been lost.

THE IMPACT OF CLIMATE CHANGE

As the world's human population increases, environmental problems are intensifying. Climate change may bring about drastic changes in the world's ecosystems and threatens to destabilise weather patterns, leading to an increase in the incidence of severe storms and droughts.

Agricultural biodiversity is our best hope for dealing with the threat that climate change poses to agriculture. Farming systems will definitely have to adapt as weather patterns change. The good news is that the most diverse farms – those that have and use the most diversity – will be better able to withstand the shocks and unpredictability of climate change.

Using agricultural biodiversity to develop crop varieties that can withstand high temperatures or that are drought-tolerant could help farmers deal with the effects of climate change, allowing them to grow their crops even as conditions get harsher.

The bad news is that climate change will have an impact on what we grow and where we grow it. The consequences could be dire for people living in the most vulnerable regions of the world. Recent research found that by 2055, more than half of the 43 crops studied – including cereals such as wheat, rye and oats – will lose land suitable for their cultivation. This loss will be particularly large in sub-Saharan Africa and the Caribbean, regions that have the least capacity to cope. And wild crop relatives – an important source of diversity – are also at risk. In 2007, scientists used computer modeling to predict the impact of climate change on the wild relatives of staple food crops. They found that, in the next 50 years, as many as 61 percent of the 51 wild peanut species analysed and 12 percent of the 108 wild potato species analysed could become extinct as a result of climate change.

CHAPTER 9

THE GUARDIANS OF DIVERSITY

All over the world there are people who dedicate their lives to safeguarding agricultural biodiversity and to using it to improve their lives and the lives of others. They are the Guardians of Diversity: individuals whose passion for diversity is helping – in small ways and large – to create a healthier, more food-secure world.

ADELAIDA CASTILLO DISPLAYS A PLAQUE NAMING HER A CHAMPION CUSTODIAN OF DIVERSITY, AN HONOUR SHE WAS AWARDED FOR HER QUINOA COLLECTION.
© A. Camacho/Bioversity International

MITSUAKI TANABE AND HIS SCULPTURE OF WILD RICE THAT HE DONATED TO THE GLOBAL CROP DIVERSITY TRUST.
© Global Crop Diversity Trust

VALERIA NEGRI WITH HER GARLIC, PEAS AND TOMATOES.
© T. Tesai

The Guardians of Diversity are farmers, researchers, writers and artists. They include Adelaida Castillo, who conserves 80 varieties of quinoa on her farm in the Peruvian Andes in memory of her son who died tragically in a motorcycle accident. They include well-known Japanese artist Mitsuaki Tanabe, who uses his art to communicate the urgent need for conserving wild rice and protecting the habitats where it grows. And they include Valeria Negri, a plant scientist at the University of Perugia in Italy, who has devoted her life to rescuing endangered Italian crop diversity.

HOW YOU CAN BE A GUARDIAN OF DIVERSITY

Talk to your parents and grandparents about the foods they used to eat when they were your age. Ask them to tell you about their food memories. Do any foods have particular meaning for them? Is it still possible to find the fruits and vegetables they used to enjoy when they were young? Write down everything they say in a notebook and compare the results with your schoolmates, who will have also interviewed their older relatives. Discuss the similarities and differences in the answers, and think of some of the reasons that people might have given different answers.

Try to find the seeds of one or more of the fruits or vegetables mentioned in the interviews with your parents and grandparents. Grow it in your garden or on a windowsill. Find out as much as you can about the plant: how it grows, how it is used, how it is eaten. With your classmates, create a book of recipes using only local, traditional plants. Eat local foods whenever you can.

Be inspired by the stories of the Guardians of Diversity at www.diversityforlife.org.

LEARN MORE

:: Agricultural Biodiversity Weblog: agro.biodiver.se
:: Bioversity International: www.bioversityinternational.org
:: The Cartagena Protocol on Biosafety: http://bch.cbd.int/protocol
:: The Global Crop Diversity Trust: www.croptrust.org
:: The International Year of the Potato: www.potato2008.org

BIODIVERSITY CONSERVATION & SUSTAINABLE DEVELOPMENT

THE NEED TO MAINTAIN NATURAL RESOURCES FOR FUTURE GENERATIONS

10

Terence Hay-Edie and Bilgi Bulus, GEF-Small Grants Programme
Dominique Bikaba, Strong Roots

What is <u>biodiversity conservation</u>? Who's involved? What do they do? How does biodiversity conservation fit into other big picture goals like <u>sustainable development</u>? What exactly is sustainable development? This chapter takes a closer look at biodiversity conservation and how it fits within the larger concept of sustainable development.

CHAPTER 10 | Biodiversity conservation & sustainable development

WHAT IS <u>SUSTAINABLE DEVELOPMENT?</u>

Humans use the planet's resources such as forests, oil and minerals. Many of these resources have accumulated or have grown over thousands or even millions of years!

The 2010 WWF Living Planet Report estimates that we'll need the equivalent of two planets by 2030 to support human populations if we continue with our current consumption patterns!

Where will we find that second planet?
What happens if we don't find it?
What alternatives are there?

Sustainable human development is about living on Earth without taking more than can be naturally replaced. It is about good health, good living conditions and long-term wealth creation for everybody. All these things must occur within the <u>carrying capacity</u> of the planet.

To understand sustainable development, think about its three pillars: "economic wealth", "social <u>equity</u>" and "environmental health"; or in other words "profit", "people" and "planet". All three are linked to each other. In other words, any development has to be not only economically sound but also beneficial to social equity and environmental health. See the box: "Defining Sustainable Development" to read the various definitions of sustainable development.

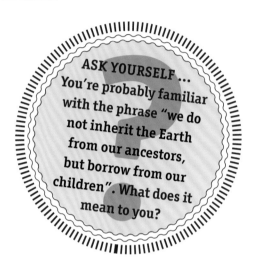

ASK YOURSELF...
You're probably familiar with the phrase "we do not inherit the Earth from our ancestors, but borrow from our children". What does it mean to you?

YOUTH AND UNITED NATIONS GLOBAL ALLIANCE

DEFINING SUSTAINABLE DEVELOPMENT

There are many ways to define sustainable development. Use the following definitions to inspire your own.

"Improving the quality of human life while living within the carrying capacity of supporting ecosystems."

The World Conservation Union (IUCN), United Nations Environment Programme (UNEP) and World Wide Fund for Nature/World Wildlife Fund (WWF)

"Sustainable development is development that meets the needs of the present without compromising the ability of future generations to meet their own needs."

United Nations in "Our Common Future, the Brundtland Report"

"Taking from the Earth only what it can provide indefinitely, thus leaving future generations no less than we have access to ourselves."

Friends of the Earth Scotland

"Sustainable development involves the simultaneous pursuit of economic prosperity, environmental quality and social equity. Companies aiming for sustainability need to perform not against a single, financial bottom line but against the triple bottom line."

World Business Council on Sustainable Development

SUSTAINABLE DEVELOPMENT

ECONOMIC GROWTH | ENVIRONMENTAL PROTECTION | SOCIAL PROGRESS

SUSTAINABLE DEVELOPMENT IS BASED ON ECONOMIC WEALTH, SOCIAL EQUITY AND ENVIRONMENTAL HEALTH.

CHAPTER 10 | Biodiversity conservation & sustainable development

CONNECTING
BIODIVERSITY CONSERVATION TO SUSTAINABLE DEVELOPMENT

What does "biodiversity" mean for people and for human development? The health of an ecosystem is closely related to the quality of life of its inhabitants. Biodiversity, as described in previous chapters, is a key component of the "environmental health" pillar of sustainable development.

Biodiversity provides people with basic <u>ecosystem goods and services</u>. It provides goods such as food, fibre and medicine, and services such as air and water purification, climate regulation, erosion control and nutrient cycling.

Biodiversity also plays an important role in economic sectors that drive development, including agriculture, forestry, fisheries and tourism. More than three billion people rely on marine and coastal biodiversity, and 1.6 billion people rely on forests and non-timber forest products (e.g. the fruits from trees) for their livelihoods. Many people depend directly on the availability of usable land, water, plants and animals to support their families. In fact, ecosystems are the base of all economies.

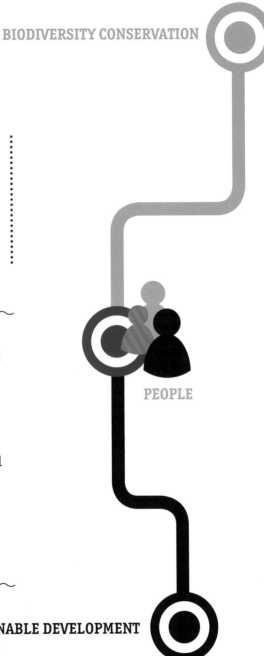

ASK YOURSELF...

It may be hard to see it at first, but when you look closely at the relationship between people and biodiversity, you will certainly recognise unsustainable behaviours. Try asking yourself some difficult questions!

- 'WHAT IF I DIDN'T OWN THIS?'
- 'DO I NEED EVERYTHING I OWN?'
- 'WHAT ARE MY REAL NEEDS?'
- 'AM I AWARE OF WHAT I EAT, HOW IT IS PRODUCED AND HOW FAR IT HAS TRAVELLED?'
- 'IS MY COMPUTER FREE OF PERSISTENT ORGANIC POLLUTANTS (POPS)?'
- 'IS MY HOUSE ENERGY-EFFICIENT?'
- 'WHAT ARE THE SOCIAL AND ENVIRONMENTAL IMPACTS OF MY LIFESTYLE?'
- 'DO I KNOW HOW TO SAVE ON ELECTRICITY AND GAS?'
- 'WHAT IS MY FAVOURITE MEANS OF TRANSPORTATION?'
- 'WHAT CAN I DO TO BE MORE SUSTAINABLE?'

CHAPTER 10 | Biodiversity conservation & sustainable development

CONSERVATION MECHANISMS

There is more than one way to conserve biodiversity. As biodiversity and its use to communities vary, so too should conservation mechanisms. Biodiversity conservation plans for a rainforest will be different from plans for a grassland or a swamp. There are different approaches involving different groups of people. There are different types of landscapes, each being used or protected for different purposes. There are different strategies and timeframes for achieving similar goals.

IN SITU AND *EX SITU* CONSERVATION

In situ and *ex situ* are the Latin words for "on-site" and "off-site". They are two different, but complementary approaches to biodiversity conservation; each plays a distinct and important role. *In situ* conservation occurs in nature, for instance in a protected area, traditional farm, nature reserve or national park. *Ex-situ* conservation occurs when a specimen of a species is set in artificial conditions such as in a zoo or a botanical garden.

In situ conservation helps to guarantee the survival of a species in its natural habitat. It is important for observing the behaviour of a species, understanding how individuals interact with other members of their species and with other species, and classifying a species as endemic (e.g. exists exclusively in a particular region), rare or under threat of extinction (see box: "*In Situ* Conservation of Great Apes"). *In situ* conservation also enables researchers to determine the distribution of a species throughout the world, to assess traditional communities' contribution to conservation, and to inform local conservation initiatives.

The IUCN, with the help of various UN agencies and governments, classifies protected areas into seven main categories:

- **Category Ia:** Integral Natural Reserve
- **Category Ib:** Wild Nature Zone
- **Category II:** National Park
- **Category III:** Natural Monument
- **Category IV:** Managed area for habitat and species
- **Category V:** Terrestrial and marine landscape
- **Category VI:** Protected area for managed natural resources.

Ex situ conservation should be used as a "last resort" or as a supplement to *in situ* conservation. *Ex situ* conservation is rarely enough to save a species from extinction. However, it is a key element for environmental and species education programmes because it provides the public with an opportunity to observe rare species from around the world in one location. If you've ever visited a zoo, animal sanctuary, botanical garden or seedbank, you've seen *ex-situ* conservation.

Different types of *ex situ* conservation have different goals and characteristics.

ASK YOURSELF …
What protected areas exist in your country? Do they house any animal species at risk of extinction?

IN SITU CONSERVATION OF GREAT APES

Did you know that all great apes are endangered species? There are four types of great apes worldwide: gorillas, chimpanzees, bonobos and orangutans. They all live in unstable and poor regions of the world such as central Africa and southeast Asia.

The Democratic Republic of the Congo (DRC) is a natural home to three of the four great apes, including the bonobo, which is endemic to the country.

MOUNTAIN GORILLAS, RWANDA.
© TKnoxB

THE YOUTH GUIDE TO BIODIVERSITY

CHAPTER 10 | Biodiversity conservation & sustainable development

DIFFERENT TYPES OF *EX SITU* CONSERVATION

Zoos focus on public education, conservation science and animal management research.

A YOUNG BOY MEETS TALINI, A POLAR BEAR CUB, AT THE DETROIT ZOO.
© David Hogg/www.flickr.com

Sanctuaries aim to protect animals and eventually release them into the wild. Most of the animals are confiscated from poachers (illegal hunters), pet traders, etc. The Pan African Sanctuaries Alliance (PASA), for example, was created to unite the African sanctuaries that had emerged in response to the deforestation, bushmeat hunting, human encroachment and disease that were decimating wild primate populations.

EASTERN COMMON CHIMPANZEES IN MAHALE MOUNTAINS NATIONAL PARK, WESTERN TANZANITA.
© G. Wales

Botanical gardens are meant for plant research, display of specimens and training.

A *VRIESEA BROMELIAD* AT THE BOTANICAL GARDEN OF MONTREAL.
© Christine Gibb

Seedbanks are more like museums; they house plant material that can be used as a source for planting if seed reserves – in cultivation or in nature – are destroyed or extinct. Seedbanks also provide researchers and breeders with crop seeds important for agriculture.

THE SVALBARD SEED VAULT IN NORWAY IS THE ULTIMATE SAFE PLACE FOR CROP DIVERSITY.
© Global Crop Diversity Trust

CHAPTER 10 | Biodiversity conservation & sustainable development

CONSERVATION ON PRODUCTIVE LANDSCAPES

While protected areas and parks remain a cornerstone for biodiversity conservation worldwide, conservation efforts are not limited to these places. Large productive landscapes with no specific conservation objective can contain lots of biodiversity while offering food, shelter and other ecosystem services for humanity.

Agricultural lands, timber forests, grasslands, rivers and marine areas are productive landscapes that are also important for biodiversity. These landscapes are managed with the aim of producing and harvesting food, timber, energy and marine resources. Even though biodiversity conservation is not the main objective, the management of these landscapes must be sensitive to biodiversity.

If it's not, resource exploitation can harm the long-term health of the ecosystem and its ability to supply food, timber, energy and other resources. This recognition has led to the promotion of sustainable agriculture, sustainable forestry, sustainable grassland management and sustainable fisheries.

MONITORING

Close monitoring of biodiversity is another important conservation practice; it involves regular checking of the overall health of ecosystems and the species living within it. The data collected from ongoing monitoring programmes can help inform management plans and improve the sustainability of activities in productive landscapes. Monitoring is especially important when the activities are carried out on an industrial scale because their impact is greater than the impact of similar activities carried out on a smaller subsistence scale by local communities.

LAW AND COMMUNITY ENFORCEMENT

Conservation mechanisms may include law or community enforcement. Biodiversity conservation officers make sure the communities relying on the site's natural resources are totally involved in conservation initiatives. Officers enforce the laws and record the details of community participation. When the laws are not respected, illegal logging, mining and bushmeat hunting erode the benefits of conservation efforts.

CERULEAN WARBLER.
© Jerry Oldenettel/www.flickr.com

COFFEE BEANS.
© Jeff Chevrier

Opposite page
A SHADE-GROWN COFFEE PLANTATION IS AN EXAMPLE OF A PRODUCTIVE LANDSCAPE THAT ALSO PROVIDES HABITAT FOR OTHER FORMS OF BIODIVERSITY. THIS COLOMBIAN COFFEE FARM PROVIDES CRITICAL WINTER HABITAT FOR DECLINING SPECIES SUCH AS THE CERULEAN WARBLER.

COFFEE PLANTATION.
© Brian Smith/American Bird Conservancy

CHAPTER 10 | Biodiversity conservation & sustainable development

TRADITIONAL KNOWLEDGE AND PRACTICES

In many cases, traditional knowledge has contributed to protecting wildlife and ecosystems and to ensuring a "natural balance". Traditional knowledge comprises of "knowledge, innovations and practices of indigenous and local communities around the world, developed from experience gained over centuries and adapted to the local culture and environment, which is transmitted orally from generation to generation", according to the Convention on Biological Diversity (CBD).

Traditional knowledge is collectively owned and takes the form of stories, songs, folklore, proverbs, cultural values, beliefs, rituals, community laws, local languages and agricultural practices, including the development of plant species and animal breeds. Article 8(j) of the CBD calls for countries to respect, preserve and maintain knowledge, innovations and practices of indigenous and local communities embodying traditional lifestyles relevant to the conservation and sustainable use of biodiversity.

Biodiversity conservation practitioners must therefore ensure that the communities relying directly on natural resources are involved in conservation initiatives, and guarantee their active participation during the whole conservation process. Some form of community engagement is essential for the success of any biodiversity conservation project.

There are also many good examples of community conserved areas around the world. These sites have been managed by communities for generations for the sustainable use of natural resources such as medicinal plants and water springs, or even for religious purposes. These sites may or may not have government protection or written management regulations. However, the community members have developed well-recognised and respected rules that are often stronger than any law and have been practised for generations. The end result is the conservation and sustainable use of resources. Some governments now legally recognise traditional practices and treat indigenous and local communities as the customary stewards of the biodiversity.

ROLE OF RESEARCH & TECHNOLOGY
FOR BIODIVERSITY CONSERVATION

Biodiversity conservation doesn't happen in a vacuum. As we've seen, it requires the participation of many different groups of people, working with various conservation mechanisms both *in situ* and *ex situ*. Biodiversity policy and conservation activities are informed, enhanced and driven by research and technology.

Researchers such as biologists, ecologists and social scientists play various roles in conservation. They identify species and their habitats, they locate areas of high ecological value, they pinpoint threats, and they propose innovative strategies and solutions. Researchers use various methods such as field surveys, observations and experiments, and technologies including remote sensing devices, data analyses, software and laboratory tests.

A RESEARCHER DOCUMENTS ILLEGAL LOGGING WITH A GLOBAL POSITIONING SYSTEM AT BUSINGIRO, BUDONGO FOREST IN UGANDA.
© Zinta Zommers, University of Oxford, UK.

CHAPTER 10 | Biodiversity conservation & sustainable development

Research results are very important for biodiversity conservation. They can be woven into community development programmes. In fact, local communities can be important contributors to biodiversity conservation research, and should be involved in all steps of the research and conservation processes. Research results are also used by conservation and development activists, journalists, government decision-makers and even businesses. The creation and application of technologies are also benefits of biodiversity conservation research. Technologies are invented, selected, evaluated, tested and applied to solve specific problems. Technologies can be transferred from rich countries to poor countries and vice versa; this process can be essential to community development. Before using any technology, however, it's critical to have a clear understanding of its characteristics so that the intervention does not harm local livelihoods, traditions, cultures or the environment.

The box: "Technological and Management Considerations for Conservation in Developing Countries" presents some of the considerations for technologies and management plans for developing countries.

KOALA, AUSTRALIA.
© Rosaria Macri

TECHNOLOGICAL AND MANAGEMENT CONSIDERATIONS FOR CONSERVATION IN DEVELOPING COUNTRIES

Biodiversity conservation in developing countries has its own set of challenges. Not only are there natural issues, such as climate change and climate variability effects, but many local communities rely on the natural resources they harvest and hunt in protected areas. In these situations, it is especially important to ensure the management plans and technologies are considerate of community needs and capacities, and are endorsed by the affected communities.

CONCLUSION

Biodiversity conservation and sustainable development are two inter-related branches focusing on social progress, economic growth and environmental protection on one side, and ecosystem conservation on the other.

Conservation includes the efforts carried out in protected areas such as national parks and community reserves, and in other areas with rich and important biodiversity where conservation is not the main focus. It is in these latter productive landscapes where sustainability is needed most. Sustainable agriculture, sustainable fisheries and sustainable management of natural resources are the main approaches for preserving these landscapes for long-term social, economic and ecological benefits.

LEARN MORE

:: The Pan African Sanctuaries Alliance: www.panafricanprimates.org
:: WWF Living Planet Report: wwf.panda.org/about_our_earth/all_publications/living_planet_report

BIODIVERSITY & PEOPLE

WE HAVE A ROLE TO PLAY: MIRRORING THE DIVERSITY OF NATURAL SYSTEMS IN DECISION-MAKING PROCESSES

Ariela Summit, Ecoagriculture Partners

Biodiversity conservation is impossible without the participation of everyone who impacts the ecosystem – from loggers who harvest forest timber, to consumers who buy food at the supermarket, to city governments who put restrictions on building in ecologically sensitive areas.

WATER AS A LIFELINE FOR PEOPLE AND ANIMALS.
© Chaitra Godbole (age 17)

CHAPTER 11 | Biodiversity & people

These stakeholders affect the diversity of the world around them, through conscious and unconscious choices they make (also see box: "Which Stakeholders Should Participate in Biodiversity Conservation?").

Inevitably, we change the ecosystems we are a part of through our presence – but we can make choices that either affirm diversity or devalue it.

For example, logging companies can choose to harvest timber in ways that sustainably thin forest cover, mimicking the actions of forest fire and making way for old growth trees.

As consumers, we can select local produce in the grocery store, supporting regionally appropriate varieties of fruits and vegetables. Local, national and international government structures that mirror the diversity of their populations are more likely to create lasting solutions to issues of food security, climate change and environmental degradation.

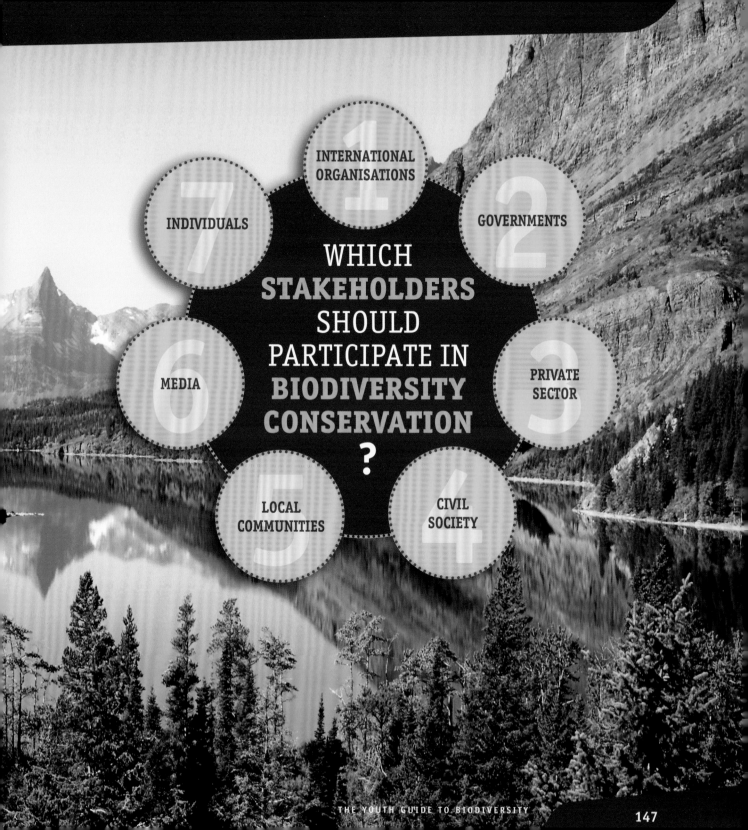

CHAPTER 11 | Biodiversity & people

WHICH STAKEHOLDERS SHOULD PARTICIPATE IN BIODIVERSITY CONSERVATION?

Stakeholders are either individuals, or representatives of a group, that have an interest, can influence or are influenced by a particular decision or action. To achieve sustainable development, the conservation of biodiversity requires the collaboration of various stakeholders including individuals, governments, private businesses, civil society, media, local communities and international organisations. Each group has an important role to play. None of these groups can stop poverty, achieve social equity, or reverse biodiversity loss alone. It is only when the groups work together that they can tackle these enormous challenges.

1 **International organisations** keep biodiversity and development on the global agenda, and determine conservation plans based on global emergencies and priorities. The United Nations (UN) works closely with governments and civil society to ensure that principles are negotiated and agreed upon, and that funding and support are provided to those who need it most.

2 **Governments** can regulate their economies so that they consider economic impacts on people and the planet. Governments develop management tools and regulations, create and implement conservation policies, and designate protected areas (e.g. national parks, community reserves, forest reserves, zoological reserves and hunting reserves).

3 **The private sector** can produce goods and services that serve people and the planet. It can provide "patient capital", a type of long-term funding available to start or grow a business with no expectation of turning a quick profit. Unlike standard business funding that often expects short- to medium-term profitability, patient capital recognises that the benefits to people and the planet can take much longer.

WHICH STAKEHOLDERS SHOULD PARTICIPATE IN BIODIVERSITY CONSERVATION?

1. INTERNATIONAL ORGANISATIONS
2. GOVERNMENTS
3. PRIVATE SECTOR
4. CIVIL SOCIETY
5. LOCAL COMMUNITIES
6. MEDIA
7. INDIVIDUALS

4 **Civil society** comprises ordinary people, citizens' groups and includes children and young people. Civil society organisations are generally non-governmental, non-profit oriented, non-military and non-individualist. They vary from large-scale professional international organisations, like the World Wide Fund for Nature/World Wildlife Fund (WWF) and the International Union for Conservation of Nature (IUCN) to local community groups like indigenous groups or your neighbourhood association. Civil society organisations represent the interests of different groups, from people using natural resources, to local communities dependent on ecosystem services, to flora and fauna and their habitats.

5 **Local communities** living in and around protected areas contribute to decision-making, and ensure that benefits arising from the use of biodiversity are equitably shared.

6 The **media** are the "mediators" between the people, governments, private sector and other actors. The media transfer information, raise awareness, and sometimes lobby for or against government or private sector decisions. Some media companies specialise in conservation issues such as the National Geographic Society in USA and the British Broadcasting Corporation (BBC) in the United Kingdom.

7 **Individual consumer choices** affect the market. Individuals should be conscious of their choices about clothing, housing, travelling, eating and other things.

If you think you don't impact the planet, think again!

Society is made of individual acts.

THE YOUTH GUIDE TO BIODIVERSITY

CHAPTER 11 | Biodiversity & people

MULTI-STAKEHOLDER
PROCESSES

> Multistakeholder processes are an important tool for creating lasting solutions for biodiversity conservation. Essentially, they are a process by which different interest groups – whether they be governments, businesses, agriculturalists or real estate developers – consult to create a plan to achieve a particular objective. Though multistakeholder processes may vary widely in scope and scale, they have certain elements in common. Typically, they are based on the democratic principles of transparency and participation.

Transparency, as used in a social science context, means that all negotiations and dialogue take place openly, information is freely shared, and participants are held responsible for their actions before, during and after the process.

The ethic of participation recognises that without all stakeholders present, solutions will not accurately address real-life pressures, and thus may not succeed.

Rural people, and particularly those who are native to the land where they live (**indigenous or aboriginal people**), are important stewards of biodiversity (see box: "Indigenous Peoples, Local Communities and Biodiversity"). Unfortunately, very often it is precisely these people who are left out of the conversation over land rights and resource management. Stakeholders who have more capital (business) or prominence (government) frequently overshadow the voices of the rural poor.

The people who have lived on the land for many generations hold invaluable storehouses of information about native varieties of plants and animals, microclimates for growing specific crops and uses of medicinal herbs. Often these same people are dependent upon these resources for survival, and have developed complex systems for maintaining the biodiversity that benefits their day-to-day lives.

150 YOUTH AND UNITED NATIONS GLOBAL ALLIANCE

Increasingly, however, rural indigenous people are tied into larger systems that profit from the large-scale destruction of these ecosystems.

When old-growth rainforest in the Amazon is burnt to make way for the cattle production to feed a growing global market for cheap beef, people who live on the land may benefit in the short term from payments for land rights or jobs. In the long term, however, they are left with the ecological consequences of land conversion, which often include polluted water, degraded fertility of the land and destruction of both plant and animal diversity. Successful strategies for biodiversity conservation must include methods of growing or gathering food in ways that do not harm, and may even benefit local biodiversity and the livelihoods of people who depend directly on the land for survival. Crafting these solutions must involve both local people, and the larger globalised markets and power structures that affect them.

> **Indigenous people are any ethnic group who inhabit a geographic region with which they have the earliest known historical connection.**

NORTH AMERICAN INDIGENOUS CHILD.
© Esperanza Sanchez Espitia

THE YOUTH GUIDE TO BIODIVERSITY

CHAPTER 11 | Biodiversity & people

INDIGENOUS PEOPLES, LOCAL COMMUNITIES AND BIODIVERSITY
John Scott, CBD

Indigenous peoples and local communities (ILCs) have a special relationship with nature in general and local plants and animals in particular, which makes them important partners of the Convention on Biological Diversity. Indigenous peoples and local communities have lived in harmony with nature and looked after the Earth's biological diversity for a long time. Their diverse cultures and languages represent much of humanity's cultural diversity. Respect for, and promotion of, the knowledge, innovations and practices of ILCs will be central to our efforts to save life on Earth.

An interesting example to illustrate the important role of indigenous peoples in maintaining biodiversity can be found in the wet tropics of far northeastern Australia. The traditional Aboriginal people of the rain forests, called the Yalanji, have practised fire management in the wet tropics for thousands of years. As a direct result of creating clearings in the jungle, grazing animals such as the kangaroo and wallaby moved into the forests from the western plain. The fire management practices of the Yalanji also encouraged the re-growth of different species of plants and fungi in these clearings.

One particular mushroom species, the main food source of a small marsupial called the bettong, grows only on the edge of such clearings.

After colonisation, many of the traditional Aboriginal peoples, including the Yalanji, were removed to church missions or government reserves, and could no longer manage their traditional lands or practise their culture. This abrupt interruption of fire management led to a decline in grazing animals found in the forest and the near extinction of the bettong. Plants living in and around the jungle clearings also fell

TASMANIAN BETTONG (BETTONGIA GAIMARDI), TASMANIA, AUSTRALIA.
© Noodle Snacks/Wikipedia

into sharp decline because many local seeds must be exposed to fire before they can germinate.

In recent years, the Yalanji have returned to their traditional lands. They are working with national park management to reintroduce fire management and the biodiversity that traditional fire management practices bring.

Source: Hill, 2004

CHAPTER 11 | Biodiversity & people

GENDER AND BIODIVERSITY CONSERVATION

In their role in the domestic sphere, and often as the primary caregiver in one-parent families, women have a central role in conserving plant genetic resources. Often, the work they do in this area is undervalued because no money changes hands in domestic transactions. Kitchen gardens maintained by wives, mothers and daughters provide an important source of micronutrients through leafy vegetables and herbs. In lean years, wild plants can be an important supplemental source of calories.

Women also hold much of the knowledge about which varieties of native species can be used for medicinal purposes, and how to prepare them safely and effectively. Lastly, women in sub-Saharan Africa, as well as indigenous societies in Latin America and the Pacific, are often directly responsible for crop production, and in this role also manage seed storage, preservation and exchange. Conserving and sustainably using biodiversity and sharing its benefits requires an understanding of and consideration for the connections between gender and biodiversity (see box: "Connections between Gender and Biodiversity"). Successful strategies for biodiversity conservation must make a special effort to include women and indigenous people. Because men and urban populations tend to have more power, education and outreach about biodiversity must be specifically targeted at those who traditionally have less of a voice and control over natural resources. To make sure that these groups are included in planning, a

A WOMAN HARVESTS PALM NUTS ON HER FARM IN AYAKOMASO, GHANA.
© Christine Gibb

154 YOUTH AND UNITED NATIONS GLOBAL ALLIANCE

ASSISTED NATURAL REGENERATION (ANR) APPROACH TO FOREST RESTORATION IN THE PHILIPPINES.
© Noel Celis

certain number of spots can be reserved for them on political committees or biodiversity project boards. If women cannot attend because of their responsibilities in the home, arrangements must be made for childcare. To include those who cannot read, educational material can be made available in other formats. In many ways, indigenous people and women have the greatest stake in preserving biodiversity, because their livelihoods directly depend on it. Thus, efforts at biodiversity conservation that also improve livelihoods have the strongest chances of success.

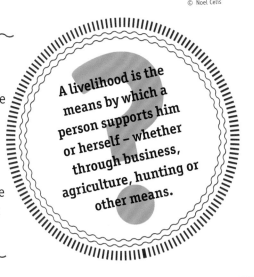

A livelihood is the means by which a person supports him or herself – whether through business, agriculture, hunting or other means.

CHAPTER 11 | Biodiversity & people

CONNECTIONS BETWEEN GENDER AND BIODIVERSITY
Marie Aminata Khan, CBD

EDUCATING *BOTH* GIRLS AND BOYS ABOUT BIODIVERSITY IS IMPORTANT. STUDENTS IN PERU LEARNING ABOUT CROP PLANTS.
© FAO/Jamie Razuri

The importance of biodiversity to individuals varies by gender. Gender refers to the social roles that men and women play and the power relations between them, which usually have a profound effect on the use and management of natural resources. Gender is not based on biological differences between women and men. Gender is shaped by culture, social relations and natural environments. Thus, depending on values, norms, customs and laws, men and women in different parts of the world take on different gender roles.

Gender roles affect economic, political, social and ecological opportunities and constraints faced by both men and women. Recognising women's roles as land and resource managers is central to the success of biodiversity policy. For example, women farmers currently account for up to 60 to 80 percent of all food production in developing countries, but gender often remains overlooked in decision-making on access to, and use of, biodiversity resources.

Just as the impact of biodiversity loss is disproportionately felt by poorer communities, there are also disparities along gender lines. Biodiversity loss affects access to education and gender equality by increasing the time spent by women and children in performing certain tasks, like collecting valuable resources such as fuel, food and water.

To conserve biodiversity, we need to understand and expose gender-differentiated biodiversity practices and gendered knowledge acquisition and usage. Various studies demonstrate that projects integrating gender dimensions generate superior results to those that don't. Gender considerations are not solely women's issues; instead, this outlook could yield advantages for whole communities and benefit both sexes.

The Millennium Development Goals emphasise clear linkages between gender equality, poverty alleviation, biodiversity conservation and sustainable development.

Such insights should be included in our outlook and approach to reversing biodiversity loss, reducing poverty and improving human well-being.

The Convention on Biological Diversity (CBD) recognises such linkages. It has developed a Gender Plan of Action that defines the CBD Secretariat's role in stimulating and facilitating efforts on national, regional and global levels to promote gender equality and to mainstream a gender perspective.

CHAPTER 11 | Biodiversity & people

LIFESTYLE CHOICES

Both rising population levels and increasing levels of consumption in the developed and developing world are responsible, to a large degree, for biodiversity loss worldwide. In the "population versus consumption" debate, some people set up a situation of extremes where they blame biodiversity loss either on rising population levels in the developing world, or on developed (mostly Western) nations who use a disproportionate share of water, fossil fuels and other natural resources.

In reality, we need to improve in both areas to save threatened plants and animals. Tools, such as the <u>environmental footprint analysis</u>, can address issues on the consumption side of the debate. This analysis is a useful tool for examining the impact that individuals have on the world around them, in terms of the resources they consume.

Choices such as eating a locally based, vegetarian diet and limiting energy usage through efficient heating and cooling systems can substantially reduce this footprint. Eating local food reduces the energy spent in transporting, processing and packaging. Eating lower on the food chain (more vegetables, legumes and grains, less meat) uses substantially less water, and also limits the amount of nutrient contamination of waterways and pollution released by livestock through methane, a gas that contributes to global warming. Energy-efficient buildings, industrial processes and transportation could also reduce the world's energy needs in 2050 by one third, according to the International Energy Agency.

TOOLS FOR BIODIVERSITY CONSERVATION

> Movements such as Slow Food International encourage a more sustainable lifestyle by celebrating regionally based cuisine and local food traditions. These foods include Reblochon cheese, which comes from the Haute Savoie region in France, or the more than 17 varieties of corn used for foods ranging from atole to tamales in Mexico.

Certification of biodiversity-friendly food informs consumers of the impacts of their food choices and provides a way to pay farmers more by producing food that also protects the environment. Biodiversity-friendly coffee, for instance, is grown in the shade under a cover of native trees, and provides habitat for birds and other wildlife.

In developing countries, ecotourism has become an important tool to preserve natural habitats while supporting local economies. Ecotourism, similar to regular tourism, involves visitors travelling to foreign countries, but is based on an ethic of environmental conservation. For instance, a tourist might stay in an energy-efficient hotel, take a safari tour to see local wildlife, and go hiking in a national park. These activities both build environmental awareness and appreciation among foreign visitors, and provide a way for local people to make a living while protecting the nature that sustains them.

Some movements to protect biodiversity are political in nature. The movements for __national sovereignty__ emphasise the value of local governance, or ensuring people from the area decide what happens with the biological resources within that area. Often, questions about who has the rights to profit from biodiversity are complicated by __land tenure__ issues, where it is unclear who actually owns the land where the resources in question are located. These issues are negotiated at a local basis, and may be influenced by international bodies such as the Convention on Biological Diversity.

Biodiversity education is an essential tool to cultivate an awareness of the value of biodiversity from an early age. This education can happen at a formal level, as integrated through school curricula, or in guides such as *The Youth Guide to Biodiversity*. See the box: "Mainstreaming Biodiversity into Education" to further explore biodiversity in formal education.

Education also happens on an informal level, though exposure to a variety of foods, cultures and environments. Such exposure tends to stimulate an awareness of diversity, and instil a creative interest in preserving it, especially when combined with larger awareness-building campaigns.

Organisations and youth groups, such as Guides and Scouts, play an important role in educating children and young people on many environmental and social issues, including biodiversity. In addition, the media can have a strong influence in raising awareness and promoting positive changes in behaviour (see the box: "Bringing the Forest to the People: The RESPECT Journey" for an example of an artistic approach to biodiversity conservation).

CHAPTER 11 | Biodiversity & people

MAINSTREAMING BIODIVERSITY INTO EDUCATION
Leslie Ann Jose-Castillo, ASEAN Centre for Biodiversity

With plant and animal extinction rates up to 100 to 1 000 times faster than the normal background rate, people can no longer afford to do nothing for biodiversity. Simply watching while the world loses key species one by one is like slowly cutting the lifeline biodiversity provides for humans – the source of food, medicine, shelter and livelihoods.

The fact that biodiversity remains unclear and intangible to many worsens the problem. People simply do not recognise the relationship between biodiversity and their well-being. It is, therefore, crucial to mainstream biodiversity into formal and non-formal learning processes.

The rationale is that learning and understanding biodiversity issues will provide people of all ages the basis for attitude change and eventual positive action conserving biodiversity.

The *Dalaw-Turo* (Visit and Teach) Programme in the Philippines illustrates how biodiversity education works in the non-formal setting.

Launched in 1989 by the Philippines' Protected Areas and Wildlife Bureau – Department of Environment and Natural Resources (PAWB-DENR) as an information, education and communication (IEC) tool for biodiversity, *Dalaw-Turo* teaches various stakeholders, particularly upland dwellers about the need to conserve biodiversity.

ABOVE AND OPPOSITE: PAWB-DENR STAFF TEACH FILIPINO STUDENTS ABOUT THE IMPORTANCE OF BIODIVERSITY CONSERVATION.
© PAWB-DENR

The programme uses street theatre, creative workshops, exhibits, games and ecological tours to stimulate creative thought and to motivate learners to act on environmental issues. Trainers from PAWB-DENR conduct school and community extension activities, train other prospective trainers, distribute IEC materials to forest occupants, local leaders, youth and school teachers. To date, *Dalaw-Turo* has trained 543 regional counterparts and brought the IEC campaign to 55 839 students in at least 460 schools in the Philippines.

In Laos, the Watershed Management and Protection Authority (WMPA) conducts the Community Outreach and Conservation Awareness programme at the Nakai Nam Theun National Protected Area. Designated WMPA staff, with the help of the village headmen, discuss with people living in the village ways to improve conservation methods in the protected area. To make the learning process interactive and informative, the teachers use games, demonstrations and role playing. Colourful and easy-to-understand posters and brochures are distributed to teach people about key species found in the area and the importance of conserving them. There is also a school education programme to teach primary students about animals, their habitats and food webs.

Children are encouraged to learn about biodiversity at an early age so that they can grow up to be protectors of the environment. These are the types of activities that must be replicated to mainstream biodiversity in education.

CHAPTER 11 | Biodiversity & people

BRINGING THE FOREST TO THE PEOPLE:
THE RESPECT JOURNEY
Christine Gibb, CBD and FAO

Images tell stories that might otherwise not be heard. Cameras record life's moments, both momentous and mundane; photos evoke emotions, questions and answers. The power of photography to tell a story is central to RESPECT, a modern-day odyssey that brings the boreal forest to the people. It is one example of the creativity and passion people bring to raising awareness about the beauty, fragility and even brutality of biodiversity.

The RESPECT journey began in Quebec, Canada and took the team of photojournalists through Ontario, Manitoba, Saskatchewan, Alberta, British Columbia and the Yukon in a small Cessna plane between 2006 and 2009. The going was tough – from turbulent weather and adverse flying conditions to unexpected interruptions and delays for equipment repair. Throughout the crossing, the team was constantly awed by the majestic landscapes of the boreal forest and its fragility; they took in breath-taking views few have had the privilege to see.

The aerial photography, however, has been viewed by millions of urbanites and tourists at several major outdoor cultural centres across Canada.

For more information about RESPECT, visit the Boreal Communications Web site at www.borealcommunciations.com

Boreal Communications

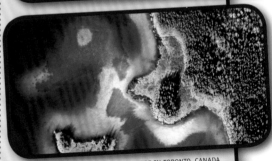

RESPECT EXHIBIT AT HARBOURFRONT CENTRE IN TORONTO, CANADA.
© Boreal Communications

ONTARIO, CANADA IN THE WINTER.
© Jim Ross/Boreal Communications

RESPECT TEAM MEMBERS: CHRIS YOUNG (PHOTOJOURNALIST), LOUISE LARIVIÈRE (CHEF-DE-MISSION) AND TOMI GRGICEVIC (VIDEOGRAPHER).
© Boreal Communications

ONTARIO LAKE AND FORESTED SHORELINE.
© Chris Young/Boreal Communications

CONCLUSION

As we've seen in this and in earlier chapters, humans are closely linked to biodiversity, through our use of biological resources and our impact on the natural world. The choices we make can have a huge impact on the current and future state of biodiversity. The next chapter takes a closer look at the decisions taken at the international level and the outcomes of these actions.

LEARN MORE

:: Boreal Communications: www.borealcommunciations.com

:: Dodds, Enayati, Hemmati and McHarry. 2002. Multi-Stakeholder Processes for Governance and Sustainability. Beyond Deadlock and Conflict. London: Earthscan.

:: FAO and gender: www.fao.org/gender/gender-home/gender-programme/gender-equity/it

:: Hill. 2004. Yalanji Warranga Kaban: Yalanji People of the Rainforest Fire Management Book. Queensland: Little Ramsay Press.

:: Howard, P. L. (ed.) 2003. Women and plants. Gender relations in biodiversity management and conservation. London: Zed Press; and New York: Palgrave Macmillan.

:: International Energy Agency: www.iea.org

:: McNeely and Scherr. 2001. Ecoagriculture: Strategies to Feed the World and Save Wild Biodiversity. Washington, DC: Future Harvest.

:: Slow Food International: www.slowfood.com

BIODIVERSITY & ACTIONS FOR CHANGE

PUTTING THE PIECES TOGETHER TO SOLVE THE BIODIVERSITY PUZZLE

12

Claudia Lewis, Plan C Initiative
Carlos L. de la Rosa, Catalina Island Conservancy

As we've seen in earlier chapters, <u>biodiversity</u> issues are complex and exist on several scales. Tackling biodiversity issues requires concerted and complementary efforts at the <u>community</u>, <u>sub-national</u>, <u>national</u>, <u>regional</u> and <u>global</u> levels.

HERMIT CRAB.
© Alex Marttunen (age 12)

CHAPTER 12 | Biodiversity and Actions for Change

In the previous chapter we looked at how individuals and groups affect biodiversity. In this chapter we will answer several important questions about the ways in which the world deals with the biodiversity challenge. For example:

- the different ways that biodiversity can be addressed at the international level;
- what actions can be taken at the national level, and how they are linked to grassroots actions;
- why <u>grassroots actions</u>, (actions undertaken by individuals or groups not associated with government) are essential to conserving biodiversity;
- how youth can help to bridge actions at the local, national and international levels.

BIODIVERSITY KNOWS **NO BORDERS**

The lines that divide countries on a map have no meaning to forest trees or roaming wildlife. These borders only become important when they turn into barriers or obstacles for biodiversity. For example, migratory herds in and around the Serengeti National Park in Tanzania and the Masai Mara National Reserve in Kenya make an extraordinary annual journey across national borders. How these two countries manage their borders and lands can mean life or death for the migrating herds of wildebeests and many other <u>species</u> that are either migratory or whose home ranges include lands on both sides of the border.

It is crucial that countries, as well as the communities within each country, reach agreements on a myriad of matters such as land use policies, exploitation of <u>natural resources</u>, <u>pollution</u> prevention, hunting regulations, water use and many other things, in order to preserve the biodiversity they share.

BLUE WILDEBEEST (CONNOCHAETES TAURINUS) IN THE NGORONGORO CRATER, TANZANIA.
© Muhammad Mahdi Karim [www.micro2macro.net]

WILDEBEEST HERDING AND FOLLOWING A FEW LEADING ZEBRA IN THE MASAI MARA, KENYA.
© T. R. Shankar Raman/Wikimedia Commons

THE MIGRATORY ROUTES OF MIGRATING WILDEBEEST TAKE LARGE HERDS ACROSS THE BORDERS OF KENYA AND TANZANIA EACH YEAR.
© www.africanhorizons.com

CHAPTER 12 | Biodiversity and Actions for Change

WE ALL DEPEND ON BIODIVERSITY,
SO WE MUST WORK TOGETHER TO MAINTAIN IT

Every country in the world depends in one way or another on biological resources. However, without careful assessment of what, how much and how often we harvest from nature, we run the risk of exhausting the planet's resources. To avoid this situation, people must work together at all levels, from local to global.

Each level of action has its own set of challenges and opportunities. What happens at one level often impacts on other levels. For example, individuals follow the conventions and regulations set in their communities, such as municipal recycling laws. Communities are bound by national legislation and laws, which regulate specific activities, such as endangered species acts or the exploitation of biodiversity. Finally, nations are bound by international agreements, such as the ones regarding the trading of wildlife species and their products. As we will see in this chapter, the laws and regulations, programmes and initiatives, treaties and informal agreements set and implemented at each of these levels can lead a country closer either to sustainability or to economic and ecological crises.

It is important to remember that whatever legislation is in place, individuals, groups or organisations can greatly influence biodiversity conservation efforts. Public opinion and campaigns can have a significant impact on policy makers and other actors, such as companies.

YUNGA BANNER AT THE WORLD SCOUT JAMBOREE IN SWEDEN.
© Maria Volodina

COMMUNITIES AND NATIONS HAVE DIFFERENT IMPACTS ON BIODIVERSITY AND RECEIVE DIFFERENT BENEFITS FROM IT

<u>Natural resource</u> use occurs at all scales; some communities exploit resources to subsist, while many others consume much more than they need for survival. Let's examine a fisheries example. At one end of the scale you have a local fishing village that only takes from the local waters what it needs to subsist, using hand-nets and other low impact practices; at the other end, you have international fishing fleets that take very high volumes of fish and other sea life, over large areas, using methods such as trolling, that have a great impact on the environment.

The impacts that these two groups have on biodiversity are quite different; the activities of the first leave a much smaller <u>footprint</u> than the second. It is not only a problem for wildlife, but also for people. Inequalities and issues of fairness need to be addressed. Cooperation and negotiation can take place at the local or at the national level, but sometimes, as in the case of fishing in international waters, global action is required.

The fisheries example is mirrored at the international level. Nations consume different amounts of resources, with some countries using a disproportionate amount of both local and global natural resources. Thus, it is necessary to hold discussions and to make agreements at the international level.

There is an uneven geographical distribution of natural resources across countries and regions. Some countries, such as the United States, possess a diverse abundance of exploitable resources, while others are not so lucky. For instance, roughly two-thirds of the Arab world depends on water sources located outside their borders. Population numbers and densities also vary widely across the world: some countries make greater demands and so impact on their natural resources more than others.

In order to have a more equitable distribution of benefits and responsibilities, and to be able to conserve biodiversity, a variety of efforts at many levels are required: establishing and enforcing agreements and treaties, implementing cooperative and assistance programmes and sharing of knowledge and technologies, are just some examples of possible action.

CHAPTER 12 | Biodiversity and Actions for Change

INTERNATIONAL ACTION

International action can occur on a regional level among several countries in a region, or on a global scale that can include many countries from several or even all continents. Such international cooperation is often critical to the success of biodiversity projects and actions (see box: "The Polar Bear Treaty").

While plants and animals do not recognise political borders between countries, people live and act within these boundaries. Thus, addressing many biodiversity issues requires the cooperation of more than one country.

INTERNATIONAL SCIENTIFIC SYMPOSIUM ON BIODIVERSITY AT FAO HQ, ITALY.
© FAO/Giulio Napolitano

For international cooperation to be effective, all of the involved countries must agree on the solutions and commit to follow the agreements. Global level efforts are indispensable when dealing with global problems, such as **climate change** and ozone depletion. These large issues often require the creation and **ratification** (or official adoption) of an international law to which all countries who are party to (or part of) it must sign (see section below).

Sometimes environmental problems are very specific to a region and/or are better addressed at the regional level. For instance, when trying to protect a species with a restricted range, such as the polar bear, or a special, fragile ecosystem and the species living within it, such as the rainforest. However, these regional approaches should still be coordinated with the broader global ones, because on our planet all things are connected. In the case of the polar bear, for instance, regional efforts to conserve it could be in vain, if the broader issue of climate change is not tackled simultaneously (see box: "The Importance of International Level Action").

THE **POLAR BEAR** TREATY

In 1973, the governments of Canada, United States, Denmark, Norway and the USSR signed a treaty that recognised the responsibilities of the countries around the North Pole for the coordination of actions to protect polar bears. The Polar Bear Treaty (Agreement on the Conservation of Polar Bears, I.L.M. 13:13-18, January 1974) commits the countries who signed it to manage polar bear populations in accordance with sound conservation practices. It prohibits hunting, killing and capturing bears, except for limited purposes and by limited methods, and commits all parties to protect the ecosystems of polar bears, especially those areas where they den and feed, as well as migration corridors.

FEMALE POLAR BEAR (URSUS MARITIMUS).
© Alan Wilson (www.naturespicsonline.com)

THE YOUTH GUIDE TO BIODIVERSITY

THE IMPORTANCE OF **INTERNATIONAL LEVEL ACTION**

Many issues related to biodiversity transcend political boundaries, so what a particular country does or does not do affects others.

Here are some examples where international level actions are important:

- International fishing regulations are needed to help prevent the over-exploitation of marine resources.
- Pollution of water bodies or overuse of water sources often requires international action, because the affected water bodies may run through more than one country. The same situation applies to air pollution.
- Alien species, pests and diseases often have impacts across national borders; their movements and effects need to be addressed at the regional and/or global levels.
- Preventing illegal wildlife trade and the smuggling of plant species requires international treaties such as the Convention on International Trade in Endangered Species of Wild Fauna and Flora (CITES) as well as coordination between many agencies and countries.
- Stabilising the Earth's changing climate will require the participation of every country in the world, especially the most industrialised ones.
- Providing funding to support the implementation of sustainability programmes in developing countries.
- International organisations provide crucial training and scientific and technical advice.
- International agreements help ensure the access to and sharing of benefits arising from the commercial use of genetic material (**bioprospecting**), and the long-term protection of biological and genetic resources.

THE UNITED NATIONS IS A HUB FOR INTERNATIONAL ACTION

The United Nations (UN) is perhaps the organisation with the largest impact and power at the global level. It has 192 member States, and conducts both regular and special meetings to address important environmental topics. Some of the most important summits include:

- The first UN Environment Conference, held in Stockholm, Sweden in 1972, leading to the establishment of the UN Environment Programme, headquartered in Nairobi, Kenya;

- The UN Conference on Environment and Development (or the 'Rio Earth Summit'), held in Rio de Janeiro, Brazil in 1992, which brought together over 179 world leaders and over 2 400 NGO representatives. It was the largest intergovernmental gathering in history, resulting in Agenda 21 (a plan of action for sustainable development), the Rio Declaration on Environment and Development, the Statement of Forest Principles, the UN Framework Convention on Climate Change (UNFCCC) and the Convention on Biological Diversity (CBD);

- The Millennium Summit held in New York, USA in 2000 where the Millennium Declaration, which includes as one of its targets the reduction of biodiversity loss, was adopted.

In addition to working with nations around the world, the UN system supports partnerships with the public and private sectors and with civil society. The UN consults NGOs and CSOs on policy and programme matters, and hosts briefings, meetings and conferences for NGO representatives.

THE UN WORLD SUMMIT ON SUSTAINABLE DEVELOPMENT TOOK PLACE IN SOUTH AFRICA.
© www.un.org

CHAPTER 12 | Biodiversity and Actions for Change

Along with other international organisations, the United Nations also provides a forum where governments can meet, discuss and agree on treaties, conventions and agreements. Such documents are known as MULTILATERAL (involving many participants), and must be signed and ratified by all participating parties to become legally binding. Once a document has been ratified, it becomes international law and replaces national laws. The ratification of these international documents (also known as 'instruments') is carried out by the congress or parliament of each country. Examples of such instruments are the 1992 Convention on Biological Diversity and the 1982 United Nations Convention on the Law of the Sea.

DIFFERENT KINDS OF INTERNATIONAL AGREEMENTS TO PROTECT BIODIVERSITY

Different kinds of international documents have different names, depending on the preferences of their signatories or the importance the instrument is meant to carry. Some of these terms can easily be interchanged; for example, an **'agreement'** might also be called a **'treaty'**.

'Protocols' or **'conventions'** are slightly less formal than treaties, because they usually contain additions or amendments to already existing treaties. Occasionally protocols or conventions contain specific obligations, like the 1997 Kyoto Protocol.

Governments, NGOs or other organisations can also enter into less formal agreements called **'declarations'**, where the parties typically declare goals which are not usually legally binding. The 1992 Rio Declaration is one example.

'Agendas' are like declarations of principles. They emerge during or as a result of international summits (meetings). They can be adopted during UN meetings, such as the General Assembly meeting in New York, or topic-specific meetings like the Rio Summit. In an agenda, countries establish common interests and priorities for a specified number of years. Agendas are basically work plans the countries set for themselves.

Finally, a **'forum'** is a less formal meeting than a summit in which one or more topics that countries would like to address can be discussed openly.

In addition to these formal global actions, there are all sorts of partnerships leading to short- and long-term collaboration among institutions, non-profit organisations and civil society.

Now let's examine some examples of international agreements and global actions.

The formation of the CBD and its work

One of the agreements adopted at the 1992 Earth Summit was the Convention on Biological Diversity (CBD), the first global agreement on the conservation and sustainable use of biological diversity. The CBD has been ratified by an overwhelming majority of countries, which are now legally committed to conserve biological diversity, use it sustainably, and share the benefits arising from the use of genetic resources equitably (fairly). The Convention offers governments and decision-makers guidance on how to deal with threats to biological diversity, and set goals, policies and general obligations. The countries are required to develop national biodiversity strategies and action plans, and to integrate them into broader national plans for environment and development. Convention-related activities undertaken by developing countries are eligible for support from the financial mechanism of the Convention: the Global Environment Facility (GEF).

The Global Environmental Facility (GEF)

Another important international initiative is the Global Environmental Facility (GEF). It was established to forge international cooperation and to finance actions to address four critical threats to the global environment: biodiversity loss, climate change, degradation of international waters and ozone depletion. It was launched in 1991 as an experimental facility, and was restructured after the 1992 Earth Summit. In 2003, two new focal areas were added: assistance for the mitigation and prevention of land degradation and persistent organic pollutants. The GEF Programme is implemented by the United Nations Development Programme (UNDP) on behalf of the World Bank and UNEP, and executed by the United Nations Office for Project Services. The GEF also has several executing agencies such as the Food and Agricultural Organization (FAO) of the United Nations and projects are supported by the UNEP, UNDP and the World Bank.

CHAPTER 12 | Biodiversity and Actions for Change

The UNDP-GEF team works with other international organisations, bilateral development agencies, national institutions, NGOS, private sector entities and academic institutions to support development projects around the world. By the end of 1999, the GEF had contributed nearly US$ 1 billion for biodiversity projects in more than 120 countries.

The UNDP, on behalf of its GEF partnership also manages two corporate accounts: the GEF National Dialogue Initiative and the GEF Small Grants Programme, which fosters environmental stewardship while helping people create and strengthen sustainable livelihoods. These small grants (under US$ 50 000) are awarded through steering committees in 73 countries. The results of three such projects are described below.

1. Testing and disseminating new technology and techniques: transforming wastes into renewable resources – Karaganda, Kazakhstan

The Karaganda Ecological Museum, an NGO in the district of Karaganda, is reducing contamination of the Nura River by providing a use for the agricultural waste that was being dumped into the river. With support from the GEF Small Grants Programme, the Museum began using agricultural waste to generate biogas and its by-products including good quality fertiliser. The Museum worked with graduate students from a local technical university to construct a biogas digester. Farmers contributed with agricultural waste and, in exchange, received biogas for cooking and lighting, and fertiliser, which increased their agricultural productivity. Not only did the project reduce the inappropriate disposal of organic wastes, it also mobilised young people to help clean up the riverbanks and spread information about the benefits of biogas.

2. Building partnerships and networks: private reserves come to the rescue of wildlife – The Cerrado biome, Brazil

According to Conservation International, the Cerrado is one of the most biologically diverse and most threatened biomes on the planet. About 70 percent of the Cerrado has suffered from human pressures of some kind, including the expansion of the Brazilian agricultural frontier for grain production and extensive cattle breeding, and from the unsustainable harvesting of woody vegetation for charcoal production. Funatura, an NGO, proposed and established four wildlife sanctuaries on private lands with the participation of other NGOs, such as the Rural Worker's Union of Formoso Municipality and the Community Association. The project is implementing mechanisms to sustain these private reserves and is disseminating the lessons learned to other landowners.

3. Developing new strategies for sustainable livelihoods in Quebrada Arroyo, Costa Rica: ecotourism for conservation and profit

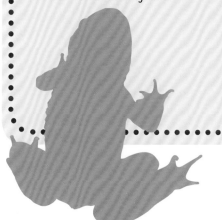

Since 1992, the GEF Small Grants Programme has supported over 30 ecotourism projects in Costa Rica. The projects are all managed by community organisations, thereby linking the protection of local biodiversity with local income generation. The village of Quebrada Arroyo, located near the Manuel Antonio National Park, one of the most visited parks in Costa Rica, is a good example of how ecotourism can protect biodiversity while generating income for a community. In 1999, a local community organisation, the Asociación de Productores de Vainilla, purchased 33 hectares that form part of the Mesoamerican Biological Corridor and then developed them for ecotourism. Today, the community receives more than 1 000 visitors per year. Women, who formerly had few economic opportunities, now earn money as tour guides. Reports indicate increases in local wildlife populations. The preservation of this area has created an important wildlife corridor connecting the Manuel Antonio National Park with the Los Santos Forest Reserve.

THE YOUTH GUIDE TO BIODIVERSITY

CHAPTER 12 / Biodiversity and Actions for Change

THE CONVENTION ON THE CONSERVATION OF MIGRATORY SPECIES OF WILD ANIMALS

Also known as CMS or the Bonn Convention, this 1993 international intergovernmental treaty sponsored by the United Nations Environment Programme (UNEP) seeks to conserve terrestrial, marine and avian migratory species across the planet. At present it includes 113 countries from Africa, Central and South America, Asia, Europe and Oceania.

The Convention encourages all the Range States to adopt global or regional agreements, which range from legally-binding treaties (called Agreements) to less formal documents, such as Memoranda of Understanding (MoUs). The box: "Connecting Biodiversity and Human Development: The Siberian Crane Wetland Project" provides an example of one such agreement.

© FAO/Giulio Napolitano

CONVENTION ON THE CONSERVATION OF MIGRATORY SPECIES OF WILD ANIMALS
Map of 116 Parties (as of 1 October 2011)
:: The Convention of Migratory Species of Wild Animals has a global reach ::

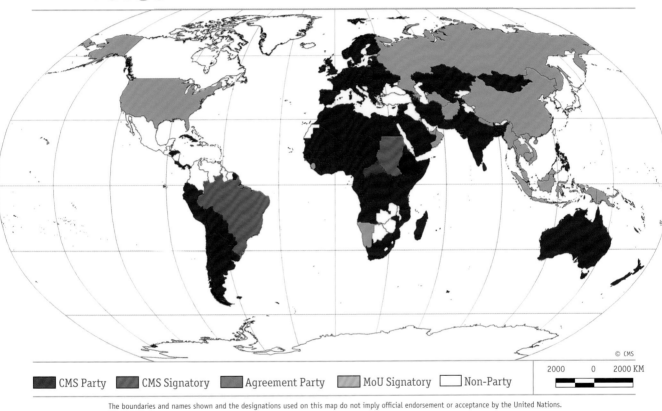

CMS Party | CMS Signatory | Agreement Party | MoU Signatory | Non-Party

The boundaries and names shown and the designations used on this map do not imply official endorsement or acceptance by the United Nations.

 See CMS video with your smart phone

THE YOUTH GUIDE TO BIODIVERSITY

CHAPTER 12 | Biodiversity and Actions for Change

CONNECTING BIODIVERSITY AND HUMAN DEVELOPMENT: THE SIBERIAN CRANE WETLAND PROJECT

The Siberian crane is the third most-endangered crane species in the world; only 3 000 to 3 500 birds remain. During its annual migration, the Siberian crane travels 5 000 km from its breeding grounds in Yakutia and western Siberia, intermediate resting and feeding places, to its wintering sites in southern China and Iran. In the last century, many of their habitats – 60 percent in Europe and 90 percent worldwide – were destroyed due to agriculture, dams, pollution and inappropriate water management, oil extraction and urban development. Additionally, unsustainable and illegal hunting led to the near *extinction* of the Western and Central Asian populations.

The CMS provided the framework for an ambitious conservation plan for the crane, covering its entire range and migration routes. The Siberian Crane Wetland Project (SCWP) is supported by UNEP's Global Environmental Facility (GEF). Government officials, experts and conservationists, such as the International Crane Foundation and Wetlands International, have worked together to use strategies to reduce hunting, to improve water management, and to mitigate the impact of climate change.

Threats to the Siberian crane and to other migratory water birds along their flyways continue to be addressed through management, monitoring, exchange of information and education of diverse audiences at local, national and international levels.

THE SIBERIAN CRANE.
© BS Thurner Hof/Wikimedia Commons

NATIONAL AND SUB-NATIONAL ACTION

Each country in the world is unique; even neighbouring countries often have different histories, customs, forms of government, needs, languages and sometimes unique ecosystems. Given this, conservation programmes must therefore be tailored to the specific conditions of a country.

For instance, in developed countries, it is often enough to buy the land and turn it into a refuge or reserve in order to protect the wildlife of a particular area.

In other countries, it is critical to secure the involvement of the local communities in the development and management of these protected areas. It often requires some form of sustainable use of the refuge, be it ecotourism, collection of seeds or plant parts, or extraction of hardwoods.

FAO GOODWILL AMBASSADOR, CARL LEWIS, TREE PLANTING IN THE DOMINICAN REPUBLIC.
© FAO/Pedro Farias-Nardi

YUNGA WORLD FOOD DAY EVENT ON BIODIVERSITY.
© FAO/Alessia Pierdomenico

THE YOUTH GUIDE TO BIODIVERSITY

CHAPTER 12 | Biodiversity and Actions for Change

WHAT CAN COUNTRIES DO TO PROTECT BIODIVERSITY?

There are many types of actions that can be implemented at national or sub-national level to tackle biodiversity issues.

These include:

3. Developing strategies, techniques and regulations to control pollution.

AFRICAN COUNTRIES LIKE KENYA, UGANDA AND TANZANIA HAVE BANNED PLASTIC BAGS IN AN EFFORT TO CURB THEIR NEGATIVE IMPACT ON WILDLIFE.
© www.dailymail.co.uk

4. Developing incentives and rewards for individuals and local communities who use sound practices in the conservation and management of their biodiversity.

COFFEE GROWERS ARE REWARDED FOR USING NATIVE SHADE SPECIES IN THEIR PLANTATIONS WHICH SUPPORT BIRDS AND OTHER WILDLIFE.
© FAO/Giuseppe Bizzarri

2. Developing plans and techniques for the management of natural resources.

MEMBERS OF THE KHARGISTAI-BAYANBURD FOREST USER GROUP, IN MONGOLIA, CLEARING TREE BRANCHES FROM THE FOREST FLOOR TO PREVENT FIRE HAZARDS.
© FAO/Sean Gallagher

5. Integrating biodiversity conservation with sustainable use, such as ecotourism.

A TOURIST PREPARING TO RIDE THE ZIPLINE AT ADVENTURE PARK IN THE PHILIPPINES. THE SUCCESS OF THE ASSISTED NATURAL REFORESTATION (ANR) PROJECT HAS HELPED THE TOURIST INDUSTRY THROUGH THE REFORESTATION OF THESE NATURAL PARKS.
© FAO/Noel Celis

1. Developing protected areas, such as parks, reserves and sanctuaries.

THE MGAHINGA GORILLA NATIONAL PARK, BWINDI IN UGANDA.
© FAO/Roberto Faidutti

Providing appropriate techniques, tools and training to communities to reduce their impact on natural resources.

TELEFOOD CHILI PROCESSING PROJECT USES SOLAR DRYERS TO PRESERVE CHILLIES.
© FAO/Giampiero Diana

Capacity-building within communities by increasing people's knowledge, skills, networking abilities and the availability of resources needed for conservation.

THE PAPUA FOREST STEWARDS INITIATIVE TRAINS LOCAL INDIGENOUS GROUPS TO SERVE AS STEWARDS OF BIODIVERSITY AND COLLABORATE WITH SCIENTISTS.
© Nomadtales/Wikimedia Commons

Providing incentives to encourage the development of wind, solar and geothermal and other more appropriate technologies and renewable energy sources.

WIND TURBINES IN NEW ZEALAND.
© Reuben Sessa

6   7   8   9   10

Passing stronger and wiser laws and regulations regarding land use, conservation easements, green corridors and urban development, to reduce habitat destruction and fragmentation.

AERIAL VIEW OF THE SHARP EDGE BETWEEN A HOUSING DEVELOPMENT AND A NATURAL AREA IN FLORIDA, USA.
© Pinellas County Government

Landscaping public urban areas to support biodiversity, and the provision of tools and incentives for urban communities to do the same.

GREEN ROOFS.
© iStock

THE YOUTH GUIDE TO BIODIVERSITY

CHAPTER 12 | Biodiversity and Actions for Change

LOCAL COMMUNITIES AND GRASSROOT ACTIONS

The success of conservation programmes and measures, whether government-initiated or not, ultimately depends on human behaviour and community action.

This is especially true of international-level programmes. No matter how clever a strategy to conserve biodiversity may be, or how strict the international treaties ratified are, unless people in the relevant areas embrace them, success is impossible or short-lived at best. The involvement of civil society can take place in many forms. Sometimes grassroots movements lead the way, other times they get involved after international and/or national organisations decide to implement a programme. In other cases, all parties may work at the same time to evaluate issues or devise and implement solutions.

LOCAL COMMUNITY INVOLVEMENT IS KEY

Governments and organisations working directly with local communities in the development of sustainable practices are more likely to be successful and have a lasting effect on the <u>conservation</u> of biological resources.

Communities are closest to these resources and are their true 'managers'. The communities are often very knowledgeable, and can provide critically important information for the development of programmes to conserve biodiversity. When communities are an integral part of the development and implementation of conservation programmes, they become empowered and have a sense of ownership of the programmes, making them more likely to care, inform and/or help other communities to follow suit. Also, by having a say in the decisions that affect biodiversity, communities can ensure that they will derive direct or indirect benefits from the conservation measures.

On the other hand, if communities are indifferent about nature, or if they lack the incentives, knowledge, resources or means to help conserve it, biodiversity pays the price.

The more informed the public are about issues and the impact of their own actions, the better they understand what constitutes harmful practices and the more willing they will be to utilise more sustainable practices, thus lessening their impact. Informed citizens can also influence environmental policies, elect politicians who will protect the environment, and remain vocal and active in keeping biodiversity issues on the agenda.

The following examples will illustrate the importance of local communities and other stakeholders in biodiversity conservation efforts.

© FAO/Riccardo Gangale

THE YOUTH GUIDE TO BIODIVERSITY

CHAPTER 12 | Biodiversity and Actions for Change

Environmental disasters caused by a lack of involvement

The Cuyahoga River in Northeast Ohio, USA, is an example of what can happen when the public doesn't get involved in the stewardship of their resources. This river was once one of the most polluted rivers in the United States, and a large portion of it was totally devoid of fish and other wildlife. The Cuyahoga became famous for being the river that caught fire, and not just once but more than a dozen times! The first fire occurred in 1868, and the largest one was in 1952. In those days, burning rivers in industrialised areas were common; rivers flowing through urban centres served as convenient sewers for industrial and human wastes. When the Cuyahoga's worst fire occurred, the citizens of Cleveland said the fire was no big deal and the chief of the fire department called it "strictly a run-of-the-mill fire"! In 1969, a fire on the river captured the attention of the Times magazine, which described it as the river that "oozes rather than flows" and in which a person "does not drown but decays".

This last fire, and the publicity it generated, finally spurred an avalanche of water pollution regulations, including the Clean Water Act, which sets limits on the amounts of pollution acceptable in all freshwater systems across the USA. Although these dramatic events occurred at a time when there were no water pollution regulations, it was the apathy and lack of involvement of the local citizens that allowed the situation to escalate to the point where it developed into an ecological catastrophe.

THE CUYAHOGA RIVER ON FIRE IN 1952.
© James Thomas, Cleveland Press Collection, Cleveland State University Library

Communities help protect threatened species in Sri Lanka

The Srinharaja Forest Reserve is Sri Lanka's last piece of fairly intact and viable tropical rainforest. The communities that depend on this forest for their subsistence have created village-level community organisations that have a say in biodiversity conservation decisions. Working in partnership with government organisations, these community CSOs actively manage and promote projects such as special zoning for various types of uses, research on selective logging and conservation of endemic flora and fauna. These organisations help change attitudes of other local residents about conservation and lead to observable results. Since the project began, for instance, illegal logging in the area has been reduced by up to 75 percent.

FARMING IN SRI LANKA.
© FAO/Ishara Kodikara

Conservation areas protected by indigenous communities in Honduras and Nicaragua

When local communities take on leadership roles and make decisions that affect their future, their actions can have national and even international conservation repercussions. The indigenous communities in the Mosquito Coast of Honduras and Nicaragua demonstrated this potential. Members of five ethnic groups (Miskito, Tawahka, Pech, Garifuna and Ladino) got together with The Nature Conservancy and its local partner NGOs to protect a corridor between two important conservation areas - Rio Plátano in Honduras and Bosawas in Nicaragua. Together, they addressed a myriad of problems: overfishing by commercial companies and other local communities, illegal harvesting of hardwoods from the forests and clear-cutting of mangroves and other lands for firewood, crop production and cattle ranching. They co-developed a long-term plan for the sustainable management of the resources on which these communities depend, including specific actions for watershed protection and sea turtle conservation. Today, these two areas the Rio Plátano (Honduras) and the Bosawas (Nicaragua) are Biosphere Reserves.

INHABITANTS OF THE BIOSPHERE RESERVES.
THREE-TOED-SLOTH.
© Stefan Laube

MANTLED HOWLER MONKEY.
© Leonardo C. Fleck

THE YOUTH GUIDE TO BIODIVERSITY

CHAPTER 12 | Biodiversity and Actions for Change

Haiti: once a lush tropical island and now an ecological disaster

The Island of Hispaniola in the Caribbean is divided into two countries. About one third of it makes up Haiti, to the west; the remaining two thirds form the Dominican Republic. Part of the border is shaped by the Libon River. But this border shows much more than the line between the two countries (see satellite image on the right): in less than a century, Haiti lost over 98 percent of its forests! As a result, over 6 000 hectares (15 000 acres) of topsoil have washed away every year, eventually leading to <u>desertification</u>, and increased pressure on the remaining land and trees. Deadly landslides, water pollution, and negative impacts on marine ecosystems are only a few of the consequences of deforestation.

Biodiversity loss has been great. USAID's Agroforestry Outreach Programme was the country's major reforestation programme in the 1980s. Local peasants planted more than 25 million trees, but for every tree planted, seven were cut. Late-coming government plans promoting alternative sources of energy for cooking to replace fuelwood and to stop deforestation have proven ineffective due to political instability and lack of funding.

This has left the communities to fend for themselves. The Dominican Republic, on the other hand, has had a more stable political climate and a better set of environmental regulations and laws. While deforestation is still an issue in the country, it has not been as devastating as in Haiti because the Dominican Republic has promoted non-extractive industries, like ecotourism, in its forests.

SATELLITE IMAGERY SHOWING DEFORESTATION IN HAITI. THE RIVER CUTTING ACROSS THE IMAGE IS THE BORDER BETWEEN HAITI (LEFT) AND THE DOMINICAN REPUBLIC (RIGHT).
© NASA USGS Landsat 7

YOUTH AND UNITED NATIONS GLOBAL ALLIANCE

Local community involvement in biodiversity protection in Mexico

The Sian Ka'an Biosphere Reserve in Mexico is home to some 2 000 people, mainly of Mayan descent. Its mission is to integrate human activities with the rich biodiversity of the region without harming the natural environment. Including local people in its management helps to maintain the balance between pure conservation and the need for sustainable use of resources by the local community.

Without the agreement and collaboration of the resident population, the area could have suffered great losses through unsustainable development.

Initiated by a presidential decree in January of 1996, it became a source of national pride when UNESCO declared it a World Heritage Site a year later.

BIOSPHERE RESERVE OF SIAN KA'AN, LOCATED IN STATE OF QUINTANA ROO, YUCATÁN PENINSULA, MEXICO.
© Tim Gage/Wikimedia Commons

Indigenous groups setting their own course in Brazil

The development of the Xingu Indigenous Park, a 2.6 million hectare (6.5 million acre) area of tropical rainforest in Brazil, is an example of a national NGO (the National Indian Foundation or FUNAI) and an international NGO (the Amazon Conservation Team or ACT) who worked with the Brazilian government's environmental agency and a coalition of 14 <u>indigenous</u> groups to achieve an unprecedented conservation milestone.

Together, they developed a series of maps describing and delineating traditional territories, fishing and hunting areas, and even sacred sites, which were incorporated into the park's management plan. The indigenous tribes participated fully in the mapping project and will be the managers of their own protected area.

Source: www.terralingua.org

BRAZIL'S XINGU INDIGENOUS PARK.
© Amazonian Conservation Team

CHAPTER 12 | Biodiversity and Actions for Change

BIODIVERSITY CONSERVATION
INVOLVES ALL STAKEHOLDERS

As you can see from this chapter, biodiversity conservation activities involve numerous actors at all levels (global to local), including:

National governments and decision-makers within ministries, agencies (e.g. ministries of environment, forestry, agriculture, fisheries and aquaculture, and regional planning) can:
- raise awareness and support education on the importance of biodiversity
- ensure that laws and policies are in place to protect biodiversity
- support collaboration and coordination between agencies at all levels (international, national, regional and local)
- support capacity building for biodiversity conservation at the local level
- ensure that local authorities have access to information
- ensure participation of all stakeholders
- provide financial resources to implement biodiversity conservation activities
- demonstrate a political commitment to implementing sustainable biodiversity management.

Local governments can:
- ensure that biodiversity considerations are included in local planning and decision-making
- promote collaboration with various stakeholders
- support local action and collaborate with NGOs, CSOs and local communities.

190   YOUTH AND UNITED NATIONS GLOBAL ALLIANCE

Universities and research institutes:
- undertake research and analysis to support improved conservation strategies and initiatives
- can provide scientific information and findings that can support awareness campaigns on informing the general public about the current state of the planet's biodiversity.

Media and celebrities can:
- highlight the perspectives of different stakeholders
- undertake independent research that gives a new view on the topic
- form, shape and influence public opinion on the importance of biodiversity conservation
- raise awareness and put pressure on decision-makers.

Environmental NGOs and CSOs can:
- act and provide support for sustainable biodiversity management at all levels
- collaborate with various stakeholders (see box: "How Non-governmental and Civil Society Organisations Help to Conserve Biodiversity").

Farmers, livestock holders, fisherfolk, land owners and local communities:
- are the local managers of biodiversity and are key to conservation actions on the ground.

The private sector can:
- provide financial resources for biodiversity initiatives
- ensure sustainable use of biodiversity products
- coordinate and collaborate with various stakeholders on biodiversity actions (see box: "Companies Can also Play a Role in Biodiversity Conservation").

The general public can:
- check and evaluate the actions of governments and other stakeholders
- demand for further action to be undertaken.

YOU! Yes, you!
yes you! Every individual can make a difference at the local, national or even international level.

CHAPTER 12 | Biodiversity and Actions for Change

HOW NON-GOVERNMENTAL AND CIVIL SOCIETY ORGANISATIONS HELP TO CONSERVE BIODIVERSITY

At the core of many biodiversity efforts are non-governmental organisations, or NGOs. An NGO is an organisation that is not part of a government; it exists for the purpose of advancing and promoting the common good, working in partnership with communities, governments and businesses to realise important goals that benefit all of society (see box: "The Cadbury Cocoa Partnership").

These organisations can work at the local, national and/or international levels. Worldwide, there are hundreds of thousands of biodiversity-related NGOs at national and local level. Examples at the international level include the World Wide Fund for Nature, The Nature Conservancy and the International Union for the Conservation of Nature (a list of these organisations is provided at the end of the chapter). NGOs are part of what the World Bank calls "Civil Society Organisations" or CSOs, which also include trade unions, faith-based organisations, indigenous people movements, foundations and many others. NGOs and CSOs sometimes work independently but often collaborate with governments.

Civil society organisations help to conserve biodiversity in various ways. They:

1) Empower local communities
2) Stimulate public awareness and action
3) Shape policy
4) Develop new strategies for sustainable livelihoods
5) Test and disseminate new and improved technologies and techniques
6) Build partnerships and networks

© FAO/Rocco Rorandelli

YOUTH AND UNITED NATIONS GLOBAL ALLIANCE

COMPANIES CAN ALSO PLAY A ROLE IN **BIODIVERSITY CONSERVATION**

The private sector can also make an important contribution to the conservation of biodiversity, for example by reducing their impact through sustainable production and business practices or by directly supporting biodiversity initiatives. Support can also be provided through public-private collaboration in which companies work with government institutions, international organisations, research centres or NGOs on biodiversity-related initiatives. Companies, especially those in sectors such as food and beverages, are highly dependent on biodiversity and ecosystem services to develop their products and undertake their operations. Hence, these biodiversity-dependent industries should have an interest in maintaining their resource base.

Public opinion and consumer choices can also influence private sector actions by putting pressure on companies to improve their social and environmental credentials.

© FAO/Ozan Ozan Kose

CHAPTER 12 | Biodiversity and Actions for Change

THE **CADBURY COCOA** PARTNERSHIP

For over 100 years, Cadbury, a world-famous chocolate and candy company, has traded cocoa with Ghana. Recently however, cocoa production has dropped significantly, which in turn has greatly reduced the incomes of cocoa farmers.

To address the problem, Cadbury, together with local government institutions, NGOs, universities and research centres and the United Nations Development Programme (UNDP), established the Cadbury Cocoa Partnership in 2005. Two key elements of the Cadbury Cocoa Partnership are the ongoing involvement of local communities and farmers in the planning and decision-making processes and the commitment to working with local organisations to turn these plans into action. Cadbury is investing £45 million pounds in the project, which will run for at least a decade.

The overall objective of the project is to directly support the economic, social and environmental sustainability of one million cocoa farmers and their communities, not only in Ghana but also in India, Indonesia and the Caribbean. To achieve this objective, the partnership is working to improve farmers' incomes by increasing yields, the quality of cocoa beans and farmers' access to fair trade schemes, providing microfinance, business support and alternative income schemes, and investing in community-led development including education projects, such as libraries and teacher training, and the building of wells for access to safe water. Cadbury has also helped produce easy-to-read illustrated newspapers containing articles about farming practices and technologies to increase cocoa production. 75 000 copies of each edition are printed and distributed to local farmers for free. Cadbury is also involved in biodiversity issues through the Earthshare programme, which it developed in partnership with the international environmental charity Earthwatch and the Nature Conservation Research Centre in Ghana.

HARVESTED MATURE CACAO FRUITS (YELLOW) AND
FRESH COCOA BEANS (WHITE).
© FAO/K. Boldt

The Earthshare programme assesses the impact of cocoa farming on biodiversity. Universities, students and volunteers work together to collect scientific information needed to help preserve biodiversity, to improve farming practices and to increase productivity.

The programme also identified additional livelihood opportunities, such as ecotourism. As a result, some farmers built simple ecotourism facilities and now earn additional income so that they are less dependent on cocoa farming.

For more information, visit:

www.innovation.cadbury.com/ourresponsibilities/cadburycocoapartnership/Pages/cadburycocoapartnership.aspx

THE YOUTH GUIDE TO BIODIVERSITY

CHAPTER 12 | Biodiversity and Actions for Change

PUTTING THE PIECES TOGETHER –
YOU CAN MAKE THE DIFFERENCE

We can all make a contribution in supporting biodiversity conservation efforts. While most of us would be content to act locally on biodiversity issues that are most accessible to us, we all have the potential to make a difference at both the national and global level.

You can also find out and contribute to local, national and international programmes and projects in a variety of ways:
- Volunteering for organisations and projects that address biodiversity issues.
- Doing internships with organisations that focus on biodiversity.
- Starting a group or club to tackle a specific issue such as invasive plants in your neighbourhood.
- Remaining informed and sharing that information with others.
- Adopting an environmentally-friendly lifestyle.
- Leading by example.

© FAO/Alessia Pierdomenico

You can also help by encouraging your government to:
- Join some of the major biodiversity protocols if your country has not already done so.
- If your country is a party to the various treaties, contact the national focal point for the Protocols to find out what is being done to implement them and how you can contribute to the process.
- Work towards strengthening national biodiversity and biosafety laws, and fostering compliance with the provisions of their protocols. Inform everyone you know on biodiversity and sustainability issues and things they can do to contribute to the conservation of biodiversity.
- Reach out to local media and write articles, including letters to the editor.

FAO and CBD have developed a number of initiatives and activities in collaboration with youth organisations, such as the World Association of Girl Guides and Girl Scouts (WAGGGS), to involve children and young people in biodiversity issues. For example, they created the biodiversity challenge badge which complements this guide book. You can download the challenge badge booklet at:
www.fao.org/climatechange/youth/68784/en

You can also get many ideas and connect with other young people at CBD's Green Wave campaign:
http://greenwave.cbd.int

Education for Sustainable Development, a UNESCO initiative, has an important youth component. On its website, youth can participate in various activities related to sustainable development and biodiversity conservation, and share ideas on how to get others involved in the discussion and solutions.
www.unescobkk.org/education/esd/esdmuralcontest

You may think:
'This is all well and good, but how do I actually start some of the above activities?'
Well, the next chapter will give you some background information, advice and ideas on how you can address biodiversity issues.

THE YOUTH GUIDE TO BIODIVERSITY

CHAPTER 12 | Biodiversity and Actions for Change

MAJOR CONVENTIONS, TREATIES, AND ORGANISATIONS
THAT WORK ON BIODIVERSITY GLOBALLY

CONVENTIONS	NICHE
Convention on Biological Diversity (CBD) and its *Cartagena Protocol on Biosafety*	The first global agreement on the conservation and sustainable use of biological diversity. The Cartagena Protocol aims to ensure the safe handling, transport and use of living modified organisms (LMOs) www.cbd.int
Convention on International Trade in Endangered Species of Wild Fauna and Flora (CITES)	Seeks to ensure that international trade of wild animals and plants does not threaten their survival. www.cites.org
Convention on the Conservation of Migratory Species of Wild Animals (CMS or Bonn Convention) and its various agreements	Aims to conserve terrestrial, marine and avian migratory species throughout their ranges on a global scale. www.cms.int
International Treaty on Plant Genetic Resources for Food and Agriculture (International Seed Treaty)	Aims to guarantee food security through the conservation, exchange and sustainable use of the world's plant genetic resources. www.planttreaty.org
United Nations Framework Convention on Climate Change (UNFCCC) and its *Kyoto Protocol*	Sets an overall framework for intergovernmental efforts to tackle the challenges posed by climate change. The Kyoto Protocol committed 55 industrialised nations to make significant cuts in the emission of greenhouse gases, such as carbon dioxide, by the year 2012. http://unfccc.int/2860.php
The United Nations Convention to Combat Desertification (UNCCD)	Tackles the problem of desertification around the world and promotes sustainable development at the community level. www.unccd.int
World Heritage Convention and its *Regional Natural Heritage Programme (RNHP)*	Promotes cooperation among nations to protect the heritage around the world which has universal value for current and future generations. The RNHP was a 4-year (2003-07) US$ 10 million programme that allocated funds to NGOs and other agencies to protect outstanding biodiversity hotspots in Southeast Asia and the Pacific. www.unesco.org/new/en/unesco

EXAMPLES OF UN-RELATED AGENDAS	
Agenda 21 of the United Nations Conference on Environment and Development (UNCED)	Comprehensive action programme that covers all areas of the environment. It specifically calls for youth involvement. www.un.org/esa/dsd/agenda21
Goal 7 of the Millennium Development Goals (MDGs): Ensure Environmental Sustainability	Adopted by world leaders in 2000 and set to be achieved by 2015, the MDGs are both global and local, tailored by each country to suit specific development needs. Goal 7 focuses on environmental sustainability and biodiversity. www.un.org/millenniumgoals
EXAMPLES OF REGIONAL CONVENTIONS and TREATIES	
Convention for the Protection and Development of the Marine Environment of the Wider Caribbean Region (WCR, also known as the Cartagena Convention) and *the Protocol Concerning Specially Protected Areas and Wildlife (SPAW)*	Legal framework for cooperative regional and national actions in the WCR for the protection and development of the marine environment. SPAW's objective is to protect rare and fragile ecosystems and habitats. www.cep.unep.org/cartagena-convention
Framework Convention for the Protection of the Marine Environment of the Caspian Sea (known as the Teheran Convention)	Protect, preserve and restore the marine environment of the Caspian Sea. http://ekh.unep.org/?q=node/2452
Polar Bear Treaty	Coordination of actions between Canada, Denmark, Norway, Russian (former USSR) and the USA to protect polar bears. www.fws.gov/laws/lawsdigest/treaty.html#POLAR
Pacific Salmon Treaty	US and Canada agreement to cooperate with regards to the management, research and enhancement of Pacific salmon stocks of mutual concern. (Amended to include the Yukon River Salmon Agreement.) www.psc.org
North Atlantic Salmon Treaty (Convention for the Conservation of Salmon in the North Atlantic Ocean -NASCO)	International organisation for the conservation and protection of Atlantic salmon. [**sedac.ciesin.columbia.edu/entri/texts/salmon.north.atlantic.1982.html**]
Northwest Atlantic Fisheries Treaty (International Convention for the Northwest Atlantic)	Investigation, protection and conservation of the Northwest Atlantic fisheries. http://treaties.un.org/doc/Publication/UNTS/Volume%201082/volume-1082-I-2053-English.pdf

CHAPTER 12 | Biodiversity and Actions for Change

EXAMPLE OF A FORUM	
United Nations Forums on Forests (UNFF)	Non-legally-binding document to promote the management, conservation and sustainable development of all types of forests. **www.un.org/esa/forests**

EXAMPLE OF INTERNATIONAL PARTNERSHIPS	
Global Invasive Species Programme (GISP) Founded by: the Centre for Agricultural Bioscience International (CABI), the Nature Conservancy, the South African National Biodiversity Institute, and the World Conservation Union	Aimed at conserving biodiversity and sustaining livelihoods by minimising the spread and impact of invasive species. **www.gisp.org/about/index.asp**
Global Taxonomy Initiative	Created to remove the taxonomic knowledge gaps of the CBD. **www.cbd.int/gti**
Census of Marine Life	Scientific initiative to assess and explain the diversity, distribution, and abundance of life in the oceans. www.coml.org

EXAMPLES OF ORGANISATIONS	MISSION
DIVERSITAS	To address the questions posed by the loss of and change in global biodiversity. **www.diversitas-international.org**
Food and Agriculture Organization of the United Nations (FAO)	To promote the conservation and sustainable use of biodiversity for food and agriculture as a means of fighting world hunger. **www.fao.org**
Global Environmental Facility (GEF)	To protect the global environment. **www.thegef.org/gef**
United Nations Environment Programme (UNEP)	To promote environmental concerns, including for biodiversity. **www.unep.org**
United Nations Educational, Scientific and Cultural Organization (UNESCO)	To contribute to the building of peace, the eradication of poverty, sustainable development and intercultural dialogue. **www.unesco.org**
World Agroforestry Centre (ICRAF)	To generate science-based knowledge about the diverse roles that trees play in agricultural landscapes. **www.worldagroforestry.org**

EXAMPLES OF NON-GOVERNMENTAL ORGANISATIONS	MISSION
Birdlife International	To conserve birds, their habitats and global biodiversity, working with people towards sustainability in the use of natural resources. www.birdlife.org
Conservation International (CI)	To protect life on Earth and to demonstrate that human societies will thrive when in balance with nature. www.conservation.org
Fauna and Flora International	To act to conserve threatened species and ecosystems worldwide, choosing solutions that are sustainable, based on sound science and take into account human needs. www.fauna-flora.org
International Union for Conservation of Nature (IUCN)	To secure living in a just and healthy environment. www.iucn.org
The Nature Conservancy	To protect ecologically important lands and water for nature and people around the world. www.nature.org
World Resources Institute (WRI)	To move human society to live in ways that protects the Earth's environment and its capacity to provide for the needs and aspirations of current and future generations. www.wri.org
World Wide Fund for Nature / World Wildlife Fund (WWF)	To preserve the diversity and abundance of life on Earth and the health of the ecological systems. www.worldwildlife.org

EXAMPLES OF OTHER CONVENTIONS, CODES OF CONDUCT AND INSTRUMENTS RELEVANT TO FOOD AND AGRICULTURE

Global Plan of Action on the Conservation and Sustainable Utilisation of Plant Genetic Resources for Food and Agriculture
www.fao.org/agriculture/crops/en

International Plant Protection Convention
www.fao.org/biodiversity/conventionsandcodes/plantprotection/en

International Treaty on Plant Genetic Resources for Food and Agriculture
www.fao.org/biodiversity/conventionsandcodes/plantgeneticresources/en

Code of Conduct for Plant Germplasm Collecting and Transfer
www.fao.org/biodiversity/conventionsandcodes/plantgermplasm/en

Code of Conduct for Responsible Fisheries
www.fao.org/biodiversity/conventionsandcodes/responsiblefisheries/en

Global Plan of Action for Animal Genetic Resources and The Interlaken Declaration
www.fao.org/docrep/010/a1404e/a1404e00.htm

BIODIVERSITY & YOU

START A PROJECT TO HELP PROTECT BIODIVERSITY YOURSELF!

Jennifer Corriero and Ping-Ya Lee, TakingITGlobal

After reading this guide and learning about the importance of biodiversity and the threats to biodiversity, get ready to take action on the issues that matter most to you. Young people around the world are leading successful projects to protect and restore the biosphere. Now it's your turn to take action: learn the six simple steps that you can take to start an action project that will help to ensure that the world's biological resources are protected for future generations.

TRANSLUCENT BUTTERFLY IN FRANCE.
© Zoe Hamelin (age 19)

CHAPTER 13 | Biodiversity & YOU

Be inspired by case studies of a youth-led reforestation initiative in Kenya and an international network of young, organic gardeners recognised at the 2009 Global Junior Challenge in Rome, Italy.

Read examples of many other projects that attendees of the Youth Symposium for Biodiversity are leading.

And **find out more** about the six simple steps that you can take to start an action project that will help to ensure that the world's biological resources are protected for future generations.

"THE WORLD IS IN OUR HANDS" BY BETTY PIN-JUNG CHEN, 1ST PLACE WINNER IN TAKINGITGLOBAL'S 2009-2010 IMPRINTS ART CONTEST.

"I hope people can love our only planet more. Think of the Earth as an egg; hold it in your hands carefully so you don't break it. The Earth is like that, we need to protect it, and love it for the next generation and the rest of our lives."

Betty Pin-jung Chen, age 13, Taiwan

SIX SIMPLE STEPS
TOWARDS CHANGE

These *Six Simple Steps Towards Change* have been adapted from the Guide to Action, created by TakingITGlobal, in consultation with young global leaders from around the world.

You can use these steps to help you to plan and execute your own biodiversity project:

1. **REFLECT AND GET INSPIRED**
2. **IDENTIFY AND GET INFORMED**
3. **LEAD AND GET OTHERS INVOLVED**
4. **GET CONNECTED**
5. **PLAN AND GET MOVING**
6. **HAVE A LASTING IMPACT**

CHAPTER 13 | Biodiversity & YOU

REFLECT & GET INSPIRED

Think about the changes you would like to see happen, whether they are in yourself, your school, your community, your country, or even the world. Who or what inspires you to take action? Seeking out sources of inspiration can give you great ideas and help you to find the strength to turn your vision into reality.

HAVE A LASTING IMPACT

Monitoring and evaluation are important parts of project management. During and at the end of your project you'll want to identify the obstacles you face and the lessons you are learning. Encouraging other youth to get involved in the issue you care about is a great way to sustain your efforts. Remember, even if you don't achieve all of your expectations, you most probably influenced others and experienced personal growth!

PLAN & GET MOVING

Now that you are equipped to take action, it's time to begin planning. Start with identifying the issue you are most interested in taking action on and one goal you can work towards. When you have your plan, stay positive and focused. Encountering obstacles is normal. You will learn a great deal more from overcoming obstacles and challenges.

IDENTIFY & GET INFORMED

What issues are you most passionate about? Learn more by gathering information about the things that interest you. By informing yourself you will be prepared to tackle the challenges that lie ahead.

LEAD & GET OTHERS INVOLVED

Being a good leader is about building on the skills you have and knowing how to leverage the strengths of others. Write down the skills that you and your team members have and see how each member of your team can use their strengths to lead in different ways. Remember that good leadership includes good teamwork!

GET CONNECTED

Networking can give you ideas, access to knowledge and experience, and help in gaining support for your project. Create a map of your networks and track your contacts.

STEPS FOR CREATING A BIODIVERSITY PROJECT.
Source: Guide to Action: Simple Steps Towards Change, TakingITGlobal, 2006

THE YOUTH GUIDE TO BIODIVERSITY

CHAPTER 13 | Biodiversity & YOU

1. REFLECT AND GET INSPIRED
Reflect on your passions

Take a moment to **REFLECT** on the biodiversity issues that matter most to you. Imagine a world of unlimited natural beauty and diversity, where humans live in harmony with the Earth's biological and natural systems. *What would that world look like?*

Think about a plant or animal <u>species</u> (see Chapter 4), or an <u>ecosystem</u> (see Chapter 5) that you want to **conserve**, **protect**, and **restore**, locally and globally.

Conserve – preserve the resilience and functions of ecosystems and biological communities by limiting the use and extraction of natural resources.

Protect – protect an ecosystem or species by campaigning to have it protected by government laws and international policies.

Restore – recreate an historic ecosystem or habitat. Improve the ecological resilience of a terrestrial or aquatic habitat. Reintroduce and encourage native species that have disappeared from an ecosystem.

ASK YOURSELF
Are there threatened plant species that you want to protect? Are there threatened animal species that you want to protect? Are there natural ecosystems near your home or abroad that you want to protect? Are there threats to biodiversity that concern you (see Chapter 2)? Are there people you know or communities abroad who are affected by threats to biodiversity?

Get Inspired

GET INSPIRED by learning about local and international biodiversity champions. Begin by reading the case studies of youth-led biodiversity projects in this chapter. You can also start to identify local biodiversity heroes in your family, neighbourhood, school or city.

Join TakingITGlobal's network of youth engaged in global issues and connect with youth leaders, organisations and projects from all over the world at **www.takingitglobal.org**.

A YOUNG STUDENT AT A LOCAL SCHOOL IN HAITI PLANTING A FRUIT TREE.
© FAO/Thony Belizaire

YOUTH AND UNITED NATIONS GLOBAL ALLIANCE

CASE STUDY: MT. KENYA YOUTH INITIATIVE FOR ECOSYSTEM RESTORATION

Sylvia Wambui Wachira, Kenya

SYLVIA WAMBUI WACHIRA (LEFT) PLANTS A TREE SEEDLING WITH A STUDENT AT KABIRU-INI SECONDARY SCHOOL, KARATINA, KENYA.
© Sylvia Wambui Wachira

The Mt. Kenya Youth Initiative for Ecosystem Restoration (MKYIER) is a volunteer-run community organisation, founded by urban and rural youth, to address deforestation in the Nyeri North district of Kenya.

Sylvia Wambui Wachira, MKYIER co-founder and programme manager, describes how she and her friends started the organisation and how they empowered a new generation of forest stewards, one school at a time:

"Schools in Kenya use firewood for cooking. Forests in Kenya are protected and it is illegal to cut trees on school properties. The schools buy wood from middlemen who source trees illegally from the forest reserve. In some school firewood piles, one can easily spot pieces of endangered indigenous trees.

So, with this basic information, my friends and I decided to fundraise from our own pockets and raised funds from our friends and relatives to start tree nurseries in schools. We gathered information from our grandparents on seeds that are good for farming and also consulted the forestry department.

We started tree nurseries in 30 schools – 19 primary schools and 11 secondary schools. We worked with existing institutions such as the Red Cross and the Scouts. And at schools where there were no groups, we created Farmers for the Future groups.

We encouraged the students to plant the tree seedlings from the nurseries around their schools. The remaining seedlings were given to the kids to plant at their homes.

We also established indigenous vegetable gardens in the school farms, which we introduced to the school groups as an activity.

We started the Mt. Kenya Youth Initiative for Ecosystem Restoration in 2006 and, now, the 10 000 trees we planted are already seven to ten metres tall."

In addition to leading MKYIER, Sylvia works as the continental coordinator of the African Youth Initiative on Climate Change (AYICC). She is also a post-graduate intern with FAO Somalia.

CHAPTER 13 | Biodiversity & YOU

CASE STUDY: ELLUMINATE FIRE & ICE ORGANIC GARDEN PROJECT

Annika Su, Taiwan

In 2006, Abel Machado School in the tiny community of Massambará, Vassouras, Brazil, participated in an online event that challenged schools to implement one activity to combat climate change in their local area. The event was organised by the non-profit global collaboration initiative, Fire & Ice, started by Elluminate, an e-learning and virtual conferencing technology provider based in Canada. In consultation with local farmers, the students at Abel Machado School created organic compost and fertilisers, and grew a wide variety of vegetables on a tiny plot of land on school property.

In 2008, Elluminate Fire & Ice invited a team from Fongsi Junior High School, Taiwan, to join this project. Annika Su was one of the students at Fongsi Junior High selected to be part of this international collaboration, along with three of her classmates. Annika and her team in Taiwan started their own organic garden at Fongsi and shared information and strategies with other schools involved in organic gardening in Mali, France, Turkey, Cuba, Solomon, Japan and Indonesia. Through blogs, virtual conferencing, PowerPoint presentations and a TakingITGlobal virtual classroom, the Elluminate Fire & Ice Organic Project engaged participating schools in a cross-cultural, educational collaboration that would gain international acclaim. In 2009, the Elluminate Fire & Ice Organic Garden Project was a finalist of the Global Junior Challenge in Rome, Italy. In Rome, Annika proudly accepted the award on behalf of her new global network of garden collaborators and friends.

When Annika first joined the Elluminate Fire & Ice Organic Garden Project in 2008, she was a "city girl" with no experience in gardening. After a full season of working to transform an empty plot at her school into a productive organic food garden, Annika developed a new appreciation for the Earth's natural processes. When asked why it is important to preserve organic agriculture methods in the face of climate change, Annika responded, "if we don't fight with nature, nature won't fight with us. That's why we should go back to working with nature."

ANNIKA SU (LEFT) AND HER CLASSMATES, RUO-CHI KONG, HUEI-CHU WU AND YU-YIN TSAI, AT FONGSI JUNIOR HIGH SCHOOL, TAIWAN, ACCEPT A 2009 GLOBAL JUNIOR CHALLENGE AWARD ON BEHALF OF THE ELLUMINATE FIRE & ICE ORGANIC GARDEN PROJECT.
© Cindea Hung

2. IDENTIFY AND GET INFORMED
Identify the issues

Refer back to your reflections on the biological resources that you would like to conserve, protect or restore. Now you can **IDENTIFY** and narrow down the biodiversity issues that are most important to you.

Which biodiversity issues are you most passionate about? What plant or animal species do you most want to protect? Is there a plant or animal species that is most important in your community?

Develop a set of questions that you want to answer. Here are some you might want to use:

- What makes this issue unique and important?
- Who is most affected by the issue, and why?
- How does this issue differ locally, nationally, regionally and globally?
- What different approaches have been taken to understand and to tackle the issue?
- Which groups are currently working on addressing the issue? Consider different sectors such as government, corporations, non-profit organisations, youth groups, United Nations agencies, etc.

Get Informed

GET INFORMED by finding resources related to issues you want to learn more about. A good place to start is with resources related to international campaigns, such as the United Nations Decade on Biodiversity (see box: "The United Nations Decade on Biodiversity"). You can go to TakingITGlobal's Issues pages to find organisations, online resources and publications for inspiration

www.tigweb.org/understand/issues.

Make a list of all the key resources you have found (organisations, publications, web sites):

1.
2.
3.
4.
5.
6.
7.

ASK YOURSELF
What more can I learn about the issues that concern me?

THE UNITED NATIONS DECADE ON BIODIVERSITY

David Ainsworth, CBD

You are part of nature's rich diversity and have the power to protect or destroy it.

Biodiversity, the variety of life on Earth, supports the living systems that provide us all with health, wealth, food, fuel and the vital services our lives depend on. We know that our actions are causing biodiversity to be lost at a greatly accelerated rate. These losses are irreversible, make our lives poorer and damage the life support systems we rely on every day. But we can prevent them.

At its 65th session, the United Nations General Assembly declared the period 2011-2020 *"the United Nations Decade on Biodiversity"*.

The decade builds on the success of the 2010 International Year of Biodiversity. The decade is a chance to think about the ways that our daily activities affect biodiversity. It's a chance to share your stories about conserving life on Earth and to inspire others to act. It's a chance to speak up and let your concern for biodiversity be heard in your town, your country and by the world leaders.

Here are a few things that you can do:

LEARN
- About biodiversity in your city, region and country.
- How your everyday actions have an impact on biodiversity, sometimes in distant ecosystems.

SPEAK
- Make your views known to government and business.
- Share your knowledge with people around you, and with the world. Post your ideas, pictures, artwork, videos and other creations on **www.facebook.com/UNBiodiversity or twitter.com/#!/UNBiodiversity.**

ACT
- Make responsible consumer choices.
- Support activities and organisations that conserve biodiversity.
- Join a local environmental group or organise your own activities that will help biodiversity.
- Be creative and find solutions to biodiversity loss.
- Continue your actions throughout the United Nations Decade on Biodiversity.

For more information, visit:
www.cbd.int/2011-2020

3. LEAD AND GET OTHERS INVOLVED
Lead your project to success

Identifying your skills and characteristics will help you **LEAD** your project to success. Start by understanding your own strengths and needs, and then consider how creating a team could help you to better achieve your goals. Helping your team members identify and leverage their own strengths and talents for the project is an important part of leadership. It is also important to ensure that all those involved are able to share in the vision of what you are trying to achieve.

Think of someone who shows strong leadership. What makes that person a good leader? Create a list of leadership qualities. Some examples are:

- **Accountable**
- **Compassionate**
- **Dedicated**
- **Fair**
- **Honest**
- **Innovative**
- **Motivational**
- **Open-minded**
- **Responsive**
- **Visionary**

List the leadership skills that you possess:
1.
2.
3.
4.
5.

List the leadership skills that you want to develop:
1.
2.
3.
4.
5.

Develop a team and get others involved

Once you have reflected on your personal leadership assets and goals, you are ready to develop a team and **GET OTHERS INVOLVED.** The box "Community Action to Improve the Environment" describes how, through the World Association of Girl Guides and Girls Scouts (WAGGGS), many girls and women have formed strong teams and have involved others. You can develop a team by starting with people you know and then expand the project to the wider community. Discuss the environmental issues in your community. How can you encourage members of the community to take part in your project to address these issues?

Name some people you already know who would want to be part of your team:
1.
2.
3.
4.
5.

What are some of the skills that your team members can contribute with?
1.
2.
3.
4.
5.

CHAPTER 13 | Biodiversity & YOU

COMMUNITY ACTION TO IMPROVE THE ENVIRONMENT
Kate Buchanan, WAGGGS

Member Organisations of the World Association of Girl Guides and Girl Scouts (WAGGGS) carry out many projects around the world.

Here are two examples of projects tackling environmental issues.

Both projects won an Olave Award for outstanding community service work at the WAGGGS World Conference in 2008.

Girl Guides of Malaysia: *"Recycling for Unity" project*

The project aim was to create awareness in the community on the importance of preserving the natural environment. This included instilling a sense of social responsibility to protect the environment, to reduce pollution, and to work towards creating a pollution-free environment at the local level.

The Girl Guides conducted a survey to assess the community's knowledge and experience in waste management and recycling. Based on the survey results, the Girl Guides visited several households and distributed information about the issues. The Girl Guides collaborated with local government, businesses and community groups to implement the project, which included distribution of recycling equipment. The project was monitored through home visits by the Girl Guides.

This was the first such project carried out in Malaysia and required significant collaboration between the private and public sectors. One unexpected benefit is that the community now has a residents' association that developed as a result of the collaboration of various groups involved in the project.

Girl Scouts of the Philippines:
regional impact project on the environment

Girl Scouts from Mindanao, an island in southern Philippines, initiated a service project on solid waste management, recycling, food production, supplementary feeding and vermiculture (the process of breeding worms to compost waste). Forty Girl Scouts transformed waste land that had become a dumpsite into an organic vegetable, ornamental and herbal community garden. The girls took part in all aspects of the project. They worked closely with local government officials, local leaders and health workers, the Department of Education and the Department of Agriculture and Girl Scout volunteers and staff.

The local community built a shed with funding from the Mayor's office. The shed served as a meeting place for the girls as well as a display area for various craft products made from recycled materials.

The health workers supervised the growing and harvesting of organic vegetables and herbs. Inspired by the girls' community action, more families in the community have constructed their own backyard compost pits and are now growing their own organic vegetables. The Girl Scouts were happy to share their time and skills. In carrying out the project, the girls broadened their attitude to community service and deepened their understanding of environmental issues.

CHAPTER 13 | Biodiversity & YOU

4. GET CONNECTED

You can also develop a team by networking and **GETTING CONNECTED** to people you have not yet met, but would like to work with. They can be associated with people who you already know, or you can try to connect with a network already working on the issues that matter to you.

You can start by attending events and conferences on biodiversity (see box: "Biodiversity Matters International Youth Symposium for Biodiversity").

List at least one event that you would like to attend:

BIODIVERSITY MATTERS
INTERNATIONAL YOUTH SYMPOSIUM FOR BIODIVERSITY
Michael Leveille and Daniel Bisaccio

"The sharing of information and collaboration with students from other nations bring us closer together and help us all to realise that we do indeed live in a global village."

Clint Monaghan,
delegate director
Second International
Youth Symposium for Biodiversity

Young people today are needed and can be meaningfully involved in protecting one small ecosystem at a time for their future and for the future of all generations.

Youth conferences are one way in which young people amplify the impact of their ideas and work.

At biodiversity symposia such as HabitatNet (Mexico, 2005) and Biodiversity Matters (Canada, 2009), youth from around the world gather to share information and strategies on youth-led projects that are making a difference.

216 YOUTH AND UNITED NATIONS GLOBAL ALLIANCE

Here are a few examples of actions that young people, like you, are undertaking:

:: High school students in Japan are protecting, breeding, and researching local owls.

:: In Ottawa, Canada, a team of elementary and middle school students are protecting an inner city marsh, and have recorded over 1 340 species.

:: A group of students from Southern India are studying and restoring a young forest sanctuary named Aranya.

:: Secondary students from Mexico and the United States are working together to protect habitats required by migratory bird species that spend part of their lives in both countries.

These types of projects are becoming a reality because youth are taking initiative and leading the way.

You can make a difference too!
Start by attending the next International Youth Symposium for Biodiversity. Join a local environmental group. Or plant a tree at your school as part of *The Green Wave* (greenwave.cbd.int).

For more information, visit: **biodiversitymatters.org**

WORLD YOUTH SYMPOSIUM ON BIODIVERSITY
JULY 5-9, 2009 • OTTAWA, ONTARIO, CANADA

THE GREEN WAVE

CHAPTER 13 | Biodiversity & YOU

5. PLAN AND GET MOVING

Develop an Action Plan

By now you have identified biodiversity issues of concern, you've learned more about the issues, and have recognised your skills and those of your team. You have also learned about the importance of networking and connecting with people who can help you to achieve your goals. You are ready to develop and implement an action **PLAN**.

Keeping in mind the issue you identified, what goal, or desirable outcome, will you work towards in your action plan? Here are some possible examples:

Conserve

- Campaign to prevent the destruction of a natural area.

- Raise awareness of a product or service that threatens biodiversity.

Protect

- Campaign to have an ecosystem recognised as a United Nations Educational Scientific and Cultural Organization (UNESCO) Biosphere Reserve.

- Get an at-risk plant or animal species on the International Union for Conservation of Nature (IUCN) Red List of Threatened Species.

Restore

- Plant native aquatic and wetland species to restore a degraded shoreline, stream bank or wetland.

- Start a tree nursery to replant an old growth forest.

© FAO/Riccardo Gangale

WRITE YOUR GOALS

Brainstorm five possible actions related to the issue you have identified. Actions are activities that will help you to achieve your goals:

1.
2.
3.
4.
5.

Develop a Mission Statement

Project Mission
Clarify what you want your project to achieve and write it down in the form of a mission statement, a short clear sentence of your purpose. For example: *Restore endangered bird habitats in local wetlands*.

Project Activities
What actions can you take to work towards achieving the mission of your project? Example: *Plant native wetland and aquatic plant species*.

Break it down
You know your mission. Now, use the sample chart below to break your project down into specific activities, resources, responsibilities and deadlines. Planning these activities in detail will ensure your project is a success. If your goal is to *restore endangered bird habitats in local wetlands*, your chart might look similar to this example:

ACTIVITY	RESOURCES	RESPONSIBILITIES	DEADLINE
Plant native wetland and aquatic species	:: Local conservation authorities :: Native plant nursery :: Grandma with her green thumb and native seed collection	*I will*: consult local conservation authorities to find out about which native species to plant *Joe will*: drive me to the native plant nursery *Grandma will*: show me how to collect seeds from native plants in her backyard	22 April, Earth Day

THE YOUTH GUIDE TO BIODIVERSITY

CHAPTER 13 | Biodiversity & YOU

Implement

Once you have finalised your plan, it is time to **GET MOVING** to implement your project. Take time to chart your progress so that you can appreciate and assess the impact of your actions. Document your project with pictures and videos. It is also a good idea to keep a project journal or blog.

Try to refer to your plan along the way, but don't expect everything to go according to the plan because many circumstances are unpredictable. You may need to revise your plan as you encounter challenges. So, remember to enjoy the entire experience as a learning process.

Raise Awareness

Create promotional materials, such as press releases and flyers, to get publicity and to let people know about your project! Word of mouth is one of the strongest marketing tools. Be enthusiastic and stay positive when you let others know how and why they should get involved. One way to promote your project is to create a project page on TakingITGlobal (takingitglobal.org) or add it to *The Green Wave* (greenwave.cbd.int) website.

Stay Motivated

Be sure to stay motivated, especially in the face of obstacles. Every challenge is an opportunity to learn. Problem-solving will require you to use your creativity to come up with innovative solutions to each challenge.

CHILDREN FROM ST. GEORGE'S SCHOOL WITH THEIR BIODIVERSITY CHALLENGE BADGE CERTIFICATES.
© FAO/Alessia Pierdomenico

6. HAVE A LASTING IMPACT

Monitoring your project throughout each stage will help you to best respond to changes that occur along the way and **have a lasting impact**. It is helpful to set out indicators or measures of success to make sure you stay on track. The more specific your indicator, the easier it will be to evaluate your achievements.

FAO AND WAGGGS TREE PLANTING CEREMONY AT FAO HQ IN ITALY.
© FAO/Alessandra Benedetti

For example:

OBJECTIVE	INDICATORS OF SUCCESS
Establish native plant nurseries on school grounds properties	:: Number of schools engaged in project :: Number of native seedlings planted :: Diversity of species planted :: Educational materials created and distributed

THE YOUTH GUIDE TO BIODIVERSITY

CHAPTER 13 | Biodiversity & YOU

CONCLUSION

Now that you have read through the Six Simple Steps Towards Change, you are ready to lead your own biodiversity action project to success. Remember that these steps are only guidelines and you may want to set your own path. There is no perfect system or path to success because each situation is unique. Every action project you start is a learning journey that will challenge you to problem-solve and to develop your own skills and talents. Don't forget to take the time to document and reflect on your progress. Keeping good records will help you to learn from your experience and it will also help you to share what you learned with other people at home and abroad. As a young biodiversity champion, you can help other youth to reflect and get inspired to start their own action projects.

Use the **Biodiversity challenge badge** to inspire you to take action: www.fao.org/climatechange/youth/68784

LEARN MORE

:: African Youth Initiative on Climate Change: www.ayicc.net
:: Biodiversity Matters: www.biodiversitymatters.org
:: Elluminate Inc.: www.elluminate.com
:: Elluminate Fire & Ice: www.elluminate.com/fire_ice/media.jsp
:: Farmers for the Future: farmersforthefuture.ning.com
:: *The Green Wave*: greenwave.cbd.int
:: *Guide to Action: Simple Steps Towards Change,* TakingITGlobal (2006): www.tigweb.org/action/guide/online.html
:: HabitatNet: www1.sprise.com/shs/habitatnet/default.htm
:: International Year of Biodiversity: www.cbd.int/2010
:: Mt. Kenya Youth Initiative for Ecosystem Restoration: www.mkyier.org
:: TakingITGlobal: www.takingitglobal.org
:: TakingITGlobal for Educators (TIGed): www.tigweb.org/tiged
:: WAGGGS: wagggs.org

CONTRIBUTORS & ORGANISATIONS

ANNEX A

LEARN MORE ABOUT THE PEOPLE WHO WROTE AND HELPED DEVELOP THIS BOOK AND ABOUT THE INSTITUTIONS WHICH HAVE BEEN INVOLVED IN ITS PREPARATION

The following annex contains information on the people and institutions which contributed to this guide, who hope that you have found the guide interesting and useful, but most of all, that you are now passionate about biodiversity and will now undertake your own actions to safeguard the world's biodiversity.

A ROOSTER IN A MARKET NEAR TOTONICAPAN, GUATEMALA.
© Lénaïg Allain-Le Drogo (age 14)

ANNEX A | Contributors & organisations

David Ainsworth is the focal person for the International Year of Biodiversity at the Secretariat of the Convention on Biological Diversity where he encourages people around the world to learn about the beauty and importance of biodiversity for our lives.

Nadine Azzu has a background in environmental management, and currently focuses on the conservation and sustainable use of biodiversity for food and agriculture. She is an Agricultural Officer at the Food and Agriculture Organization of the United Nations.

Daniel J. Bisaccio is Director of science education (Master of Arts in Teaching) and Clinical Professor in education, in Brown University's Education Department. He is the founder of HabitatNet and a lead organiser of both the 2005 and 2009 International Youth Symposia on Biodiversity.

Dominique Bikaba has a degree in Rural Development and a specialisation in Regional Planning. He is the Executive Director of the Strong Roots. Previously, he coordinated the Pole Pole Foundation, which won the UNDP-Equator Initiative finalist prize in 2006.

Kate Buchanan is Programme Development Coordinator for the World Association of Girl Guides and Girl Scouts (WAGGGS). She develops educational programmes for girls and young women, including activities to help them learn about the UN's Millennium Development Goals and how to set up community action projects.

Zeynep Bilgi Bulus has studied business administration and agro-food economy. Her professional career has been spent as a nature conservationist, first in the Turkish Society for the Protection of Nature, and later in the GEF-SGP hosted by the UNDP. She currently lives on a farm in Turkey and continues to give volunteer and professional consultancies to civil society and international organisations.

David Coates works on inland waters biodiversity at the Secretariat of the Convention on Biological Diversity. He works in particular on land and water management issues and the role of inland water ecosystems in supporting sustainable development.

Jennifer Corriero is a social entrepreneur and youth engagement strategy consultant with a Masters in Environmental Studies from York University. She is co-founder and Executive Director of TakingITGlobal, and has been recognised by the World Economic Forum as a Young Global Leader.

Carlos L. de la Rosa is the Chief Conservation and Education Officer for the Catalina Island Conservancy, in Southern California, USA. He holds a doctorate in aquatic ecology and has worked for over 20 years in conservation issues in Latin America and North America. Currently he oversees many projects and initiatives in biodiversity conservation and environmental education.

Amanda Dobson is a graduate of John Cabot University, Rome. She is currently a Programme Assistant at the Global Crop Diversity Trust. She interned for the Diversity for Life campaign during the summer of 2009.

Maria Vinje Dodson is a Communications and Development Officer at the Global Crop Diversity Trust.

Cary Fowler has worked in the conservation and use of crop diversity for more than 30 years. He is currently the Executive Director of the Global Crop Diversity Trust and chair of the Advisory Council of the Svalbard Global Seed Vault.

Christine Gibb is a Consultant for the Secretariat of the Convention on Biological Diversity and for the Food and Agriculture Organization of the United Nations. Her current projects focus on youth and biodiversity issues.

Jacqueline Grekin is a Programme Assistant at the Secretariat of the Convention on Biological Diversity. Her work includes inland waters biodiversity, marine and coastal biodiversity, island biodiversity and agricultural biodiversity.

Caroline Hattam is an Environmental Economist at Plymouth Marine Laboratory. She works on projects that aim to encourage the sustainable use and management of the marine environment.

Terence Hay-Edie joined the GEF-SGP to provide technical support to the country programmes in the areas of biodiversity, protected areas and projects relating to indigenous peoples. Prior to UNDP, he worked with the UNESCO World Heritage Centre and Man and the Biosphere programmes.

Saadia Iqbal was the Former Editor of the World Bank's Youthink! Web site. Her projects included writing and producing multimedia content, as well as outreach and collaboration with partner organisations. She is currently writing a children's book.

Leslie Ann Jose-Castillo was the Development Communication Expert at the ASEAN Centre for Biodiversity. Her work focused on communication, education and public awareness, programme promotion and media relations.

ANNEX A | Contributors & organisations

The late Marie Aminata Khan was the Gender Programme Officer for the Secretariat on Biological Diversity. She also worked on communications and outreach activities within the division of Outreach and Major Groups.

Conor Kretsch is an environmental scientist specialising in ecosystems conservation and management and the links between the environment and human well-being. He is the Executive Director at the COHAB Initiative Secretariat.

Ping-Ya Lee was the Coordinator of the Tread Lightly programme, a climate change education and youth engagement initiative developed by TakingITGlobal. She worked on developing new content for the programme and also promoted the free tools and resources to teachers and schools. She is currently a master of landscape architecture candidate at the University of Toronto.

Michael Leveille is a science teacher at St-Laurent Academy Elementary and Junior High School in Ottawa, Canada. He is the Executive Director of the Second International Youth Symposium for Biodiversity and the founder of the Macoun Marsh Biodiversity Project.

Claudia Lewis is a conservation biologist and psychologist by training, and an environmental educator by trade. She currently works as an environmental consultant and as the Executive Director of Plan C Initiative, in Florida. She has devoted over 20 years of her career to educate and empower individuals of all ages, but especially communities, to conserve biodiversity and live more sustainably.

Charlotte Lusty is a scientist at the Global Crop Diversity Trust, working with national genebanks and international partners to help regenerate unique collections and secure them in long-term conservation.

Ulrika Nilsson works for the Cartagena Protocol on Biosafety at the Secretariat of the Convention on Biological Diversity as an Associate Information Officer. She works on public awareness and participation, including media and outreach in the field of biosafety.

Kieran Noonan-Mooney works at the Secretariat of the Convention on Biological Diversity where he helped to prepare the third edition of the Global Biodiversity Outlook (GBO-3).

Kathryn Pintus has a background in zoology and conservation, and now works for the IUCN's Species Programme, where she contributes to the advancement of biodiversity conservation through communications.

Neil Pratt is the Senior Environmental Affairs Officer at the Secretariat of the Convention on Biological Diversity. He oversees outreach, communication and education issues with each of the major stakeholders, including children and young people.

Ruth Raymond is the Communications Manager of Programmes and Regions at Bioversity International. She has more than 20 years of experience in raising awareness about the value of agricultural biodiversity.

John Scott is a descendant of the Iningai people of central Queensland, North Eastern Australia. He is currently the Programme Officer for article 8(j) (Traditional Knowledge) at the Secretariat of the Convention on Biological Diversity and the focal point for indigenous peoples and local communities. The focus of his work is on the legal protection of traditional knowledge.

Reuben Sessa is a Programme Officer at FAO developing and coordinating programmes on climate change. He is also FAO focal point for youth, coordinator of the YUNGA initiative and member of the Inter-Agency Network on Youth Development.

Junko Shimura works on taxonomy and invasive alien species at the Secretariat of the Convention on Biological Diversity. She works on capacity development for countries to identify, monitor and manage biodiversity, including the control of pathways for the introduction of invasive alien species.

Ariela Summit is completing a Master's Degree in Urban and Regional Planning at UCLA, with a focus in environment and community economic development. She formerly worked at Ecoagriculture Partners in Washington DC coordinating the US programme and managing outreach efforts.

Giulia Tiddens works for YUNGA coordinating activities related to social media and events such as the biodiversity World Food Day event with other 450 children. She is also involved in the preparation of materials and activities related to biodiversity and forests.

Tamara van 't Wout works for FAO on projects strengthening young people's and countries' capacities for disaster risk reduction and climate change adaptation.

Jaime Webbe works for the Secretariat of the Convention on Biological Diversity. She is responsible for the biodiversity of dry and subhumid lands and for the interactions between biodiversity and climate change.

ANNEX A | Contributors & organisations

www.aseanbiodiversity.org
The ASEAN Centre for Biodiversity is an intergovernmental regional centre of excellence which facilitates cooperation and coordination among ASEAN Member States and with relevant national governments, regional and international organisations on the conservation and sustainable use of biodiversity and the fair and equitable sharing of benefits arising from the use of such biodiversity.

www.bioversityinternational.org
Bioversity International is the world's largest international research organisation dedicated solely to the conservation and use of agricultural biodiversity.

www.biodiversitymatters.org
At the Second International Youth Symposium for Biodiversity, 100 students and their chaperones gathered in Ottawa, Canada in July 2009. The International Youth Accord for Biodiversity was initiated, and was presented at COP-10 in Japan in 2010. The Macoun Marsh Project, an award-winning youth biodiversity project of St-Laurent Academy, hosted the Symposium.

www.brown.edu
At Brown University, students study education from a variety of disciplinary perspectives, including anthropology, economics, history, political science, psychology, biological/ physical sciences, and sociology. The faculty teaches a wide array of undergraduate and graduate courses, and conducts research on important educational issues.

www.catalinaconservancy.org
The Catalina Island Conservancy is a land trust located on Catalina Island, in Southern California. Part of the California Channel Islands and the California Floristic Province Hot Spot for Biodiversity, the Conservancy stewards its 42 000 acres of land through a balance of conservation, education and recreational activities. A non-profit and public charity organisation, the Conservancy partners with state, national and international organisations in developing solutions for the most pressing biodiversity issues.

www.cohabnet.org
The COHAB (Co-operation on Health and Biodiversity) Initiative is an international programme of work addressing the gaps in knowledge, awareness and action on the links between biodiversity and human health. COHAB works around the world to promote greater awareness of the importance of biodiversity to human health and well-being, and supports projects to improve community health through conservation.

www.cbd.int and bch.cbd.int/protocol
The Convention on Biological Diversity is an international agreement that commits governments to maintaining the world's ecological sustainability through conservation of biodiversity, sustainable use of its components, and the fair and equitable sharing of the benefits arising from the use of genetic resources. The Cartagena Protocol on Biosafety is one of the key international agreements contributing to environmental conservation and sustainable development by reducing the potential negative effects that living modified organisms may pose to biodiversity.

www.ecoagriculture.org
EcoAgriculture Partners strives for a world where current agricultural lands are increasingly managed as ecoagriculture landscapes to achieve three complementary goals: to enhance rural livelihoods; conserve biodiversity; and sustainably produce crops, livestock, fish, and forest products. As a non-profit organisation, EcoAgriculture Partners helps to scale up successful ecoagriculture approaches by catalysing strategic connections, dialogue, and joint action among key actors at local, national and international levels.

www.fao.org
The Food and Agriculture Organization of the United Nations (FAO) leads international efforts to defeat hunger. FAO acts as a neutral forum where all nations meet as equals to negotiate agreements and debate policy. Forums on biodiversity hosted by FAO include the Commission on Genetic Resources for Food and Agriculture (CGRFA), the Governing Body of the International Treaty on Plant Genetic Resources for Food and Agriculture and the Commission on Phytosanitary Measures (which governs the International Plant Protection Convention - IPPC). FAO is also a source of knowledge and information, helping countries to modernize and improve agriculture, forestry and fisheries practices and ensure good nutrition for all.

ANNEX A | Contributors & organisations

www.croptrust.org
The Global Crop Diversity Trust's mission is to ensure the conservation and availability of crop diversity for food security worldwide.

www.iucn.org
The International Union for Conservation of Nature (IUCN) helps the world find pragmatic solutions to our most pressing environment and development challenges by supporting scientific research; managing field projects all over the world; and bringing governments, NGOs, the UN, international conventions and companies together to develop policy, laws and best practises. IUCN is the world's oldest and largest global environmental network with more than 1 000 government and NGO members, and almost 11 000 volunteer scientists and experts in 160 countries.

www.plancinitiative.org
Plan C Initiative is an organisation based in Florida, USA, whose mission is to empower local urban communities and municipalities to develop landscapes that support people and wildlife, utilising strategies that enhance community connections and ecological function. The organisation endeavours to create a paradigm shift that leads to ecological landscaping becoming the prevailing approach to urban landscaping. The vision is that natural areas in Florida become interconnected via a large urban ecological landscape that supports increased biodiversity and provides a variety of services to humans and wildlife.

www.pml.ac.uk
Plymouth Marine Laboratory (PML) is an independent, impartial provider of scientific research, contract services and advice for the marine environment. PML's work focuses on understanding how marine ecosystems function and reducing uncertainty about the complex processes and structures that sustain life in the seas and their role in the Earth system.

www.sgp.undp.org
Established in 1992, the year of the Rio Earth Summit, the GEF Small Grants Programme (SGP) provides financial and technical support to community-based projects that conserve and restore the environment while enhancing people's well-being and livelihoods. SGP demonstrates that community action can maintain the fine balance between human needs and environmental imperatives in five focal areas: biodiversity conservation, mitigation of climate change, combating land degradation, phasing out of persistent organic pollutants, and protection of international waters.

www.strongrootscongo.org
Strong Roots is a local organisation based at the Kahuzi-Biega National Park (KBNP) in eastern Democratic Republic of Congo (DRC). It involves indigenous and local communities in the long-term preservation of the park through sustainable development projects. Strong Roots has programmes on environmental education, carving, reforestation, crop production and food security, sustainable land management, health, conservation and small business activities.

www.tigweb.org
TakingITGlobal is a non-profit organisation with the aim of fostering cross-cultural dialogue, strengthening the capacity of youth as leaders, and increasing awareness and involvement in global issues through the use of technology.

www.wagggsworld.org
The World Association of Girl Guides and Girl Scouts (WAGGGS) is a worldwide movement providing non-formal education where girls and young women develop leadership and life skills through self-development, challenge and adventure. Girl Guides and Girl Scouts learn by doing. The Association brings together Girl Guiding and Girl Scouting Associations from 145 countries reaching 10 million members around the globe.

www.youthink.worldbank.org
Youthink! is the World Bank's Web site for youth. It informs and engages young people on development issues.

www.yunga.org
YUNGA: the Youth and United Nations Global Alliance is a partnership of different UN agencies, civil society organisations and other groups related to children and youth. The objective of the alliance is to create resources and activities to educate and engage children and young people in activities of key environmental and social concern at the national and international levels. YUNGA also seeks to empower children and young people to have a greater role in society, raise awareness and be active agents of change.

PLEASE NOTE THAT THE INVOLVEMENT OF AN INSTITUTION OR INDIVIDUAL DOES NOT IMPLY ITS OR THEIR ENDORSEMENT OR AGREEMENT WITH THE CONTENT OF THIS GUIDE.

LIST OF SPECIES

SCIENTIFIC NAMES

ANNEX B

In addition to one or more common names, every internationally recognised species has a unique scientific name. It consists of two names, usually Latin or Greek words, that are always italicised (or underlined if the name is written by hand). The first name is the genus (generic name), and begins with a capital letter; the second is the species (specific name), and is written in lower-case.

Scientific names are given according to a taxonomic classification system called 'binomial nomenclature'. The system was first introduced in the eighteenth century by a Swedish botanist named Carl von Linné (sometimes called Carolus Linneaus).

MOSQUITO ON RED PETAL.
© iStock

ANNEX B | List of Species

The binomial nomenclature system has several benefits:
- it's simple (only two names)
- it's clear (a single species could have many common names in several languages, but it only has **one** scientific name)
- it's stable over time (with some exceptions)
- it's used extensively around the world.

The common and scientific names of the species described in this publication are listed in the following table.

COMMON NAME	SCIENTIFIC NAME
African elephant	*Loxodonta africana*
African forest elephant	*Loxodonta cyclotis*
Alpine ibex	*Capra ibex*
Amazon River dolphin, Boto	*Inia geoffrensis*
Auroch	*Bos primigenius*
Australian honey possum	*Tarsipes spenserae*
Black and white jumping spider	*Phidippus audax*
Black and white ruffed lemur	*Lemur varius*
Black bear	*Ursus americanus*
Blue whale	*Balaenoptera musculus*
Bornean orangutan	*Pongo pygmaeus*
Bottlenose dolphin	*Tursiops truncatus*
Brown pelican	*Pelecanus occidentalis*
Burrowing bettong	*Bettongia lesueur*
Cane toad	*Bufo marinus*
Cerulean warbler	*Dendroica cerulean*
Chum salmon	*Oncorhynchus keta*
Colossal squid	*Mesonychoteuthis hamiltoni*
Cycad species	*Encephalartos brevifoliolatus*
Dodo	*Raphus cucullatus*
Domestic cow	*Bos taurus*
Eastern common chimpanzee	*Pan troglodytes schweinfurthii*
Eastern lowland gorilla	*Gorilla beringei*
Eastern spinner dolphin	*Stenella longirostris*
Everglade kite	*Rostrhamus sociabilis*
Firefly squid	*Watasenia scintillans*
Florida panther	*Puma concolor coryi*
Fly agaric mushroom	*Amanita muscaria*
Ganges River dolphin, Susu	*Platanista gangetica*
Giant clam	*Tridacna gigas*
Giant panda	*Ailuropoda melanoleuca*
Gile monster lizard	*Heloderma suspectum*
Giraffe	*Giraffa camelopardalis*

COMMON NAME	SCIENTIFIC NAME
Gnu, blue wildebeest	*Connochaetes taurinus*
Great crested tern	*Sterna bergii*
Green lacewing	*Chrysoperla rufilabris*
Green turtle	*Chelonia mydas*
Hippopotamus	*Hippopotamus amphibious*
Hula painted frog	*Discoglossus nigriventer*
Human	*Homo sapiens*
Indus River dolphin	*Platanista minor*
Irrawaddy River dolphin, Mekong River dolphin	*Orcaella brevirostris*
Jaguar	*Panthera onca*
King penguin	*Aptenodytes patagonicus*
Long-horned bee	*Tetraloniella spp.*
Night blooming jasmine	*Cestrum nocturnum*
Pacific yew tree	*Taxus brevifolia*
Pig-footed bandicoot	*Chaeropus ecaudatus*
Plains zebra	*Equus quagga*
Polar bear	*Ursus maritimus*
Potato (cultivated)	*Solanum tuberosum*
Potato (wild)	*Solanum megistacrolobum*
Red wriggler worm	*Eisenia foetida*
Ringed seal	*Pusa hispida*
Sea otter	*Enhydra lutris*
Siberian crane	*Grus leucogeranus*
Stem rust	*Puccinia graminis*
Striped lychnis moth	*Shargacucullia lychnitis*
Thomson's gazelle	*Eudorcas thomsonii*
Tiger	*Panthera tigris*
Traveller's tree, traveller's palm	*Ravenala madagascariensis*
Tucuxi	*Sotalia fluviatilis*
Wood stork	*Mycteria americana*
Woolly-stalked begonia	*Begonia eiromischa*
Yangtze River dolphin, Baiji	*Lipotes vexillifer*

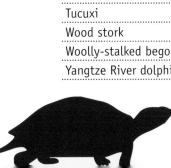

Glossary

GLOSSARY

Action: an activity that will help you to achieve your goal(s).

Action plan: a strategy that helps you to break your project down into specific activities, resources, responsibilities and deadlines. Planning these activities in detail will ensure the success of your project.

Agricultural biodiversity: the components of biodiversity important for agriculture.

Agricultural productivity: the ratio of agricultural outputs to agricultural inputs. When productivity is high the farmer harvests much more than he or she puts into the land.

Agro-ecosystem: an ecosystem where there is agricultural activity. Agro-ecosystems include land used for crops, pasture and livestock; the adjacent uncultivated land that supports other vegetation and wildlife; and the associated atmosphere, underlying soils, groundwater and drainage networks.

Amphibians: a large group of animals that usually have moist skins and live in or in association with freshwater - including frogs, toads, newts and salamanders. Most have eggs without shells, which are laid or develop in water or moist environments.

Anther: the male part of a flower that produces pollen.

Aquaculture: the cultivation of marine or freshwater animals (e.g. fish, molluscs and crustaceans) and aquatic plants.

Background rate: the normal extinction rate, based on the fossil record before humans became a major contributor to extinctions.

Ballast water: ballast refers to anything used by ships to help them create stability. Sea water is the usual form of ballast.

Biodiversity: the variety of life on Earth, at each of the genetic, species and ecosystem levels, and the relationships between them.

Biodiversity hotspot: an area especially rich in plant and animal life, but in grave threat of being destroyed. To be recognised as a biodiversity hotspot, the area must: have at least 1 500 endemic species of vascular plants, and have lost at least 70 percent of its original habitat.

Biofuel: a fuel made from living or recently living biological materials such as marine algae, maize or sugarcane.

Biomagnification: the accumulation of substances in organisms that increase in concentration up the food chain, as smaller organisms are eaten by larger organisms.

Biomass: in ecology, biomass is the mass of living organisms in an ecosystem at a given time.

Bioprospecting: the study and potential commercialisation of useful plant and animal species.

Biosafety: efforts to reduce possible risks from modern biotechnology and its products, including measures to ensure the safe transfer, handling and use of living modified organisms created through modern biotechnology.

Brackish water: water that is saltier than fresh water, but not as salty as seawater. It occurs in transitional areas between rivers and the sea, such as estuaries and mangrove swamps.

Breeding: the production of plant or animal offspring. Breeding can refer to the intentional breeding of specially selected parents by farmers or researchers.

Carrying capacity: the population size of a species that the environment can sustain indefinitely, given available food, space, light, water and nutrients.

Cell: the basic building block of life. All organisms are made up of one or more cells.

Civil society organisation (CSO): an organisation that is not part of a government. In addition to NGOs, the term CSO includes trade unions, faith-based organisations, indigenous people movements, foundations and many others.

Climate change: a direct driver of biodiversity loss. It is a change in the overall state of the Earth's climate caused by both natural and human causes such as the build-up of greenhouse gases, like carbon dioxide, in the Earth's atmosphere.

Clone: a genetically identical copy of a cell or individual.

Conservation: changing needs or habits with the aim of maintaining the health of the natural world, including land, water, biodiversity and energy.

Complementary interests: when different parties share the same interest in the same parcel of land (e.g. when members of a community share common rights to grazing land, etc.).

Competing interests: when different parties contest the same interests in the same parcel (e.g. when two parties independently claim rights to exclusive use of a parcel of agricultural land).

Crop wild relative: a non-cultivated species that is more or less closely related to a crop species (usually in the same genus). Crop wild relatives are not normally harvested for food but they can occur in farmers' fields (e.g. as a weed or a component of pasture or grazing lands), and are important sources of diversity for crop improvement.

Dead zone: coastal sea areas where water oxygen levels have dropped too low to support marine life. They often result from the build-up of nutrients, usually carried from inland farming areas where fertilisers wash into watercourses. The nutrients promote the growth of phytoplankton that die and decompose on the seabed, using up the oxygen in the water and threatening fisheries, livelihoods and tourism.

Desertification: the degradation of land in arid and semi-arid areas causing the deterioration of the ecosystem or the loss of agricultural production.

Direct driver: a direct cause of biodiversity loss. The five main ones are: habitat loss and fragmentation, climate change, invasive alien species, pollution and overexploitation of resources / unsustainable use.

Drylands: dry and subhumid lands that include everything from deserts to savannas to Mediterranean landscapes.

Ecology: the scientific study of the relationships between and among organisms, and of all aspects of their environment.

Ecosystem: the combined physical and biological components of an environment, and their interactions. An ecosystem is relatively self-contained and is defined by the types of organisms found there and their interactions (e.g. forest, grassland, lake).

Ecosystem goods and services: the benefits that the environment, including humans, obtains from ecosystems. These benefits include cleaning the air and water, and providing food and materials to build houses. There are four types of ecosystem services: provisioning, regulating, cultural and supporting.

Endemic: a species that is native to a particular area or environment and not found naturally anywhere else.

Environmental footprint analysis: a useful tool for examining the impact that individuals have on the world around them, in terms of the resources they consume.

Equity: something that is fair and just.

Eutrophication: a process whereby water bodies receive excess nutrients that stimulate excessive plant growth. In turn, this enhanced plant growth reduces dissolved oxygen in the water and can cause other organisms to die.

Evaporation: the process whereby a liquid turns into a gas.

Evolution: the gradual process of genetic change that occurs in populations of organisms, eventually leading to new species. It requires natural selection, diversity, inheritability and time.

Ex situ conservation: off-site conservation in which plants or animals are removed from their natural habitat and placed in a new location such as a zoo or seed bank.

Extinction: the state whereby no live individuals of a species remain.

Forage crops: plants that are eaten by livestock animals. Examples of forage plants are different kinds of grasses, or herbaceous legumes such as clover or alfalfa.

Fragmentation: a direct driver of biodiversity loss. It is a process whereby parts of a habitat become separated from one another because of changes in a landscape. Fragmentation makes it difficult for species to move throughout a habitat, and poses a major challenge for species requiring large tracts of land.

GLOSSARY

Gender: the social roles that men and women play and the power relations between them, which usually have a profound effect on the use and management of natural resources.

Gene: a section of DNA which encodes information about the characteristics of the organism; it is a unit of heredity and is passed down from parent to offspring.

Genebanks: institutions where genetic diversity is conserved and documented for use by farmers and researchers. The most specialist institutions are able to keep species and varieties in good health in storage for several decades.

Gene pool: the total number of genes belonging to every individual in an interbreeding population.

Genetic diversity (or variability): the variation and richness of genes in a population or species.

Genetic erosion: the loss of genes from a population or the loss of species from an ecosystem.

Genus: a low-level rank used to classify living and fossil organisms.

Goal: a desirable outcome.

Grassroots action: an action undertaken by individuals or groups not associated with the government.

Habitat: the local environment in which an organism is usually found.

Habitat loss: a direct driver of biodiversity loss. It occurs when natural environments are transformed or modified to serve human needs.

Homogeneous: similar. Homogeneous groups are the same or very similar.

Horticultural crops: crops including fruits, berries, nuts and vegetables. Horticulture is the practice of cultivating plants.

Hypoxic: a place is said to by hypoxic when the level of oxygen available in water to living organisms is below the level that these organisms need to survive.

Indicator: a measure of success to make sure you stay on track of your goal.

Indigenous people: any ethnic group which inhabits a geographic region with which they have the earliest known historical connection.

Inheritability: the ability to transfer resources, in this case traits or genes, from parent to offspring.

In situ conservation: on-site conservation in which plants or animals are protected in their natural habitats, either by protecting or cleaning up the habitat, or by defending the species from disease, competitors and predators.

Invasive alien species (IAS): a direct driver of biodiversity loss. IAS are species that have spread outside of their natural habitat and threaten biodiversity in the new area. IAS may also cause economic or environmental damage, or adversely affect human health.

Invertebrate: an animal that does not have a backbone.

Land tenure: the relationship, whether legally or customarily defined, among people, as individuals or groups, with respect to land. Land tenure constitutes a web of intersecting interests.

Livelihood: the means by which a person supports him or herself – whether through business, agriculture, hunting or other means.

Living modified organism (LMO): an organism produced by modern biotechnology in which scientists have taken a single gene from a plant, animal or micro-organism, and inserted it into another organism. LMOs are commonly known as genetically modified organisms (GMOs).

Marine biodiversity hotspots: these are areas of high species and habitat richness that include representative, rare and threatened features.

Micro-organism: a living thing too small to be seen by the human eye alone, but that can be seen through a microscope. In ecosystems, micro-organisms help in recycling nutrients.

Mission statement: a short clear sentence of your purpose.

Morphology: the study of the form and structure of individual organisms (what the organism looks like).

Mulching: in an agro-ecosystem that produces crops, mulching is the practice of leaving a cover of organic material on the soil. Mulching provides material for worms to digest and recycle into nutrients for plants, but it also has other benefits for the environment. For example, it prevents loss of water from evaporation, helps to reduce erosion, and helps to suppress weed growth.

Multilateral: involving a large number of parties.

National sovereignty: the power of a state to do everything necessary to govern itself, such as making, executing, and applying laws; imposing and collecting taxes; making war and peace; and forming treaties or engaging in commerce with foreign nations.

Natural resource: something from nature that can be used to make something else; farmers need natural resources, such as land, air, water and sunlight, to grow food.

Natural selection: the survival of animals and plants that adapt successfully to their environment and produce healthy offspring enabling the transfer of their genes and traits.

Networking: building a team and getting connected to people who can help you to achieve your goal(s).

Non-governmental organisation (NGO): an organisation that is not part of a government. It exists for the purpose of advancing and promoting the common good, working in partnerships with communities, governments and businesses in realising important goals that benefit all of society. These organisations can work at the local, national and/or international levels.

Ocean acidification: a decrease in ocean pH due to increased levels of atmospheric carbon dioxide dissolving in seawater.

Omnivore: an organism that eats a wide variety of foods, including foods of plant and animal origin.

Orchard: an example of an agro-ecosystem that is used for growing fruit or nut trees for consumption and/or commercialisation.

Organism: an individual living creature such as a spider, walnut tree or human.

Overexploitation: a direct driver of biodiversity loss. It happens when biodiversity is removed faster than it can be replenished. It is also called "unsustainable use".

Overriding interests: when a sovereign power (e.g. a nation or community) has the power to allocate or reallocate land.

Overlapping interests: when several parties are allocated different rights to the same parcel of land (e.g. one party may have lease rights, another may have a right of way, etc.).

Patient capital: a type of long-term funding available to start or grow a business with no expectation of turning a quick profit.

Peatland: an area with a thick, naturally accumulated organic layer of peat on its surface. Peat is made of dead and partially decomposed plant remains that have accumulated on the spot under waterlogged conditions. Peatlands include wetlands such as moors, bogs, fens, mires, swamp forests or permafrost tundra. They are found in all biomes, particularly the boreal, temperate and tropical areas of the planet. Peatlands are important for carbon storage, water retention, biodiversity, agriculture, forestry and fisheries.

Persistent organic pollutants (POPs): organic compounds that do not easily break down through chemical, biological and photolytic processes. They accumulate in the environment and can be hazardous to human health and the environment.

Phenology: the timing of biological events such as flowering and fruiting of plants.

Photophore: a special body part found on deep sea creatures that is bioluminescent (produces light).

GLOSSARY

Phylogenetics: the study of evolutionary relationships, using genetics to look at how closely related different species are.

Phylum: a technical term used in the classification of living creatures, and refers to a broad group of related organisms.

Phytoplankton: microscopic aquatic plants that drift in the upper parts of the ocean.

Pollinator: an animal that carries pollen from one seed plant to another, unwittingly helping the plant to reproduce. Common pollinators include bees, butterflies, moths, birds and bats.

Pollution: a direct driver of biodiversity loss. It occurs where contaminants, such as chemicals, energy, noise, heat and light are introduced into an environment and destabilise or harm the ecosystem.

Protected areas: places that receive protection because of their environmental or cultural value.

Ratification: official adoption, in this case, of an international agreement.

Renewable resource: a natural resource that can replenish itself.

Reptiles: snakes, lizards, crocodiles, turtles and tortoises, etc. Some are terrestrial (land-living), others live on both land and in water, some exclusively in water (e.g. freshwater turtles). Most produce eggs with shells which are laid and develop out of water.

Services: see ecosystem services.

Sexual dimorphism: the occurrence of physical differences between individuals of different sexes (not including primary sexual characteristics) but of the same species. Male and female peacocks, for instance, look very different: the males have large, colourful tail feathers that the females lack.

Species: a group of similar organisms which are able to breed together and produce healthy, fertile offspring (offspring that are able to produce young).

Stakeholders: those that have an interest in a particular decision, either as individuals or as representatives of a group. It includes people who influence a decision, or can influence it, as well as those affected by it.

Stigma: the female part of a flower which receives pollen.

Subspecies: a taxonomic rank below species. Subspecies of a species will be different from one another in some way, but the differences are not so great as to consider them separate species. Geographic isolation of populations of a species may result in the evolution of certain traits, which in turn may lead to the formation of a subspecies.

Sustainable development: development that meets the needs of the present without compromising the ability of future generations to meet their own needs.

Symbiosis: a relationship which provides a mutual benefit. A symbiotic relationship is a mutually beneficial relationship between two species.

Taxonomy: the science of naming, describing and classifying organisms.

Terrestrial biodiversity: all of the animals and plants and micro-organisms that live on land, and also land habitats, such as forests, deserts and wetlands.

Trait: a characteristic or distinguishing feature that identifies an organism, for instance curly hair or tallness. In agriculture, important traits include those that affect a plant's yield or resistance to disease. Some traits are heritable and others are not.

Transparency: when all negotiation and dialogue take place openly, information is freely shared, and participants are held responsible for their actions before, during, and after the process.

Transpiration: a process in which plants return water to the atmosphere.

Unsustainable use: a direct driver of biodiversity loss. It happens when biodiversity is removed faster than it can be replenished. It is also called "overexploitation".

Vector: any living or non-living carrier that transports living organisms intentionally or unintentionally.

Vertebrates: animals that have a backbone or a spinal cord. Some examples of vertebrates are mammals, birds, sharks and reptiles.

Water footprint: the total volume of fresh water that is used to produce the goods and services consumed by an individual, business or nation.

Wetland: an area of land covered either permanently or temporarily with water, usually shallow, covered by plants (including trees) which grow out of the water or mixed with areas of open water.

Zooplankton: microscopic aquatic animals that drift in the upper parts of the ocean.

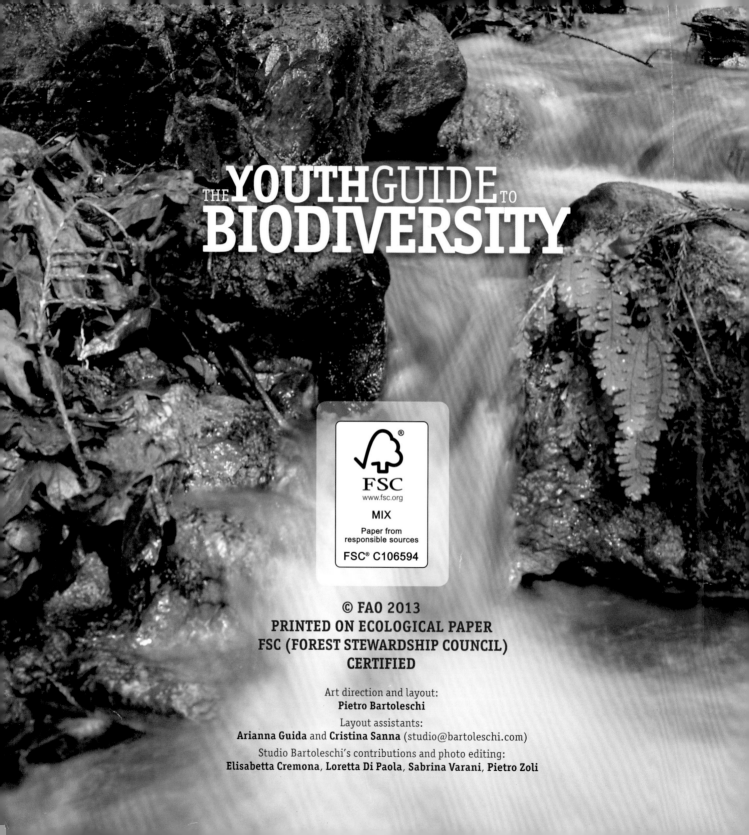

THE YOUTH GUIDE TO BIODIVERSITY

© FAO 2013
PRINTED ON ECOLOGICAL PAPER
FSC (FOREST STEWARDSHIP COUNCIL)
CERTIFIED

Art direction and layout:
Pietro Bartoleschi
Layout assistants:
Arianna Guida and **Cristina Sanna** (studio@bartoleschi.com)
Studio Bartoleschi's contributions and photo editing:
Elisabetta Cremona, **Loretta Di Paola**, **Sabrina Varani**, **Pietro Zoli**